Niels Benson

Organic CMOS technology by dielectric interface engineering

Niels Benson

Organic CMOS technology by dielectric interface engineering

Chemically functionalized dielectrics for the control of OFET polarity / charge carrier transport properties

Südwestdeutscher Verlag für Hochschulschriften

Impressum/Imprint (nur für Deutschland/ only for Germany)
Bibliografische Information der Deutschen Nationalbibliothek: Die Deutsche Nationalbibliothek verzeichnet diese Publikation in der Deutschen Nationalbibliografie; detaillierte bibliografische Daten sind im Internet über http://dnb.d-nb.de abrufbar.
Alle in diesem Buch genannten Marken und Produktnamen unterliegen warenzeichen-, marken- oder patentrechtlichem Schutz bzw. sind Warenzeichen oder eingetragene Warenzeichen der jeweiligen Inhaber. Die Wiedergabe von Marken, Produktnamen, Gebrauchsnamen, Handelsnamen, Warenbezeichnungen u.s.w. in diesem Werk berechtigt auch ohne besondere Kennzeichnung nicht zu der Annahme, dass solche Namen im Sinne der Warenzeichen- und Markenschutzgesetzgebung als frei zu betrachten wären und daher von jedermann benutzt werden dürften.

Verlag: Südwestdeutscher Verlag für Hochschulschriften Aktiengesellschaft & Co. KG
Dudweiler Landstr. 99, 66123 Saarbrücken, Deutschland
Telefon +49 681 37 20 271-1, Telefax +49 681 37 20 271-0, Email: info@svh-verlag.de
Zugl.: Darmstadt, TU, Diss., 2009

Herstellung in Deutschland:
Schaltungsdienst Lange o.H.G., Zehrensdorfer Str. 11, D-12277 Berlin
Books on Demand GmbH, Gutenbergring 53, D-22848 Norderstedt
Reha GmbH, Dudweiler Landstr. 99, D- 66123 Saarbrücken
ISBN: 978-3-8381-0765-3

Imprint (only for USA, GB)
Bibliographic information published by the Deutsche Nationalbibliothek: The Deutsche Nationalbibliothek lists this publication in the Deutsche Nationalbibliografie; detailed bibliographic data are available in the Internet at http://dnb.d-nb.de.
Any brand names and product names mentioned in this book are subject to trademark, brand or patent protection and are trademarks or registered trademarks of their respective holders. The use of brand names, product names, common names, trade names, product descriptions etc. even without
a particular marking in this works is in no way to be construed to mean that such names may be regarded as unrestricted in respect of trademark and brand protection legislation and could thus be used by anyone.

Publisher:
Südwestdeutscher Verlag für Hochschulschriften Aktiengesellschaft & Co. KG
Dudweiler Landstr. 99, 66123 Saarbrücken, Germany
Phone +49 681 37 20 271-1, Fax +49 681 37 20 271-0, Email: info@svh-verlag.de

Copyright © 2008 Südwestdeutscher Verlag für Hochschulschriften Aktiengesellschaft & Co. KG and licensors
All rights reserved. Saarbrücken 2008

Produced in USA and UK by:
Lightning Source Inc., 1246 Heil Quaker Blvd., La Vergne, TN 37086, USA
Lightning Source UK Ltd., Chapter House, Pitfield, Kiln Farm, Milton Keynes, MK11 3LW, GB
BookSurge, 7290 B. Investment Drive, North Charleston, SC 29418, USA
ISBN: 978-3-8381-0765-3

Contents

1	**Introduction**	**3**
2	**Theoretical fundamentals**	**7**
	2.1 Organic semiconductors	7
	2.2 Transistors	12
	2.2.1 Organic field effect transistors	13
	2.2.2 Influence of the dielectric interface on charge carrier transport in organic field effect transistors	18
	2.3 Electrets	20
3	**Experimental Framework**	**23**
	3.1 Materials	23
	3.2 Sample preparation	25
	3.2.1 Thin film deposition	25
	3.2.2 Sample structure	26
	3.2.3 Electrical characterization and sample conditioning	27
	3.3 Thin film characterization	30
	3.3.1 Atomic force microscopy	30
	3.3.2 Water contact angle measurement	30
	3.3.3 Layer thickness determination	30
	3.3.4 Photoelectron spectroscopy	31
4	**Ambipolar/Unipolar OFET charge carrier transport**	**37**
	4.1 OFET polarity in dependence of the source/drain metalization	38
	4.2 Influence of different dielectrics on OFET charge carrier transport	40
5	**OFET dielectric interface engineering**	**43**
	5.1 Ca modified Silicon dioxide	43
	5.1.1 n-type transport in dependence of a Ca modified SiO_2 interface	44
	5.1.2 XPS interface analysis	46
	5.1.3 Correlation between transistor performance and PES data	51
	5.1.4 Influence of thermal and electrical stress on OFET transport properties	51
	5.2 UV modified Polymethylmetacrylate	57

	5.2.1	Introduction of charge carrier traps on PMMA	57
	5.2.2	Influence of UV modified PMMA gate dielectrics on OFET transport properties	65
5.3	Applications		73

6 OFET threshold tuning by the use of an electret   77

7 Summary   81

Nomenclature   83

List of Figures   87

List of Tables   91

Literature   93

Acknowledgements   103

A List of publications   105

B Additional PES spectra   106

C Development in OFET mobility   109

D Process Parameters   111
    D.1 Thin film deposition from solution 111
    D.2 Thin film deposition by PVD 112

Chapter 1

Introduction

Due to the extensive research and a positive development in organic electronics during the last 25 years, manufacturers such as Pioneer, Kodak, Siemens as well as Sony were able to bring products to the market incorporating organic light emitting diode (OLED) displays. Simply stated, an OLED consists of an organic semiconductor, with an optical bandgap in the visible to the near infrared range of the electromagnetic radiation spectrum, sandwiched between a reflective and a transparent electrode. By injecting both electrons and holes into the organic layer, light emission is achieved. A second major branch in organic electronics development deals with organic field effect transistors (OFETs), the scope of interest for the work at hand. In general, research and commercial interest in organic electronics is motivated by the promise of low-cost processing on flexible substrates thus making the technology very attractive for price effective consumer applications. However, the short, one or two year life for current products, e.g., car radios, digital cameras, cell phones and even television sets, indicate component reliability issues in the production process or the presence of other life cycle problems. This implies, that presumably low-cost electronic products are brought to market, based primarily upon marketing aspects of the new technology. Nevertheless, the stage is now set for a consequent introduction of organic electronics in the market. In particular upcoming products, such as radio frequency identification (RFID) tags or a new e-Reader featuring a rollout display based on an OFET active matrix and others, will demonstrate, whether we are a step closer to this goal. This dissertation is dedicated to advances in related organic CMOS technology.

Research on organic solids with respect to photoconductivity can be traced back to the beginnings of the 20th century, where Pochettino (1906) [1] and Volmer (1913) [2] conducted first experiments on anthracene. By the 1940s, actual semiconducting properties of π-conjugated small molecules or polymers were demonstrated [3] and first potential applications in xerography [4, 5], photovoltaics as well as the electroluminescence [6, 7] and the field effect [8] in organic solids were investigated throughout the following three decades. However, the first organic field effect transistors were not demonstrated until the mid 1980s [9, 10]. These OFETs were implemented, using organic dyes such as merocyanine, or polymer organic semiconductors such as polythiophene. From its beginning, the OFET performance ($\mu \approx 10^{-6} \frac{cm^2}{Vs}, \frac{On}{Off} \approx 10^{2-3}$) has significantly improved over the years, even rivaling today's electrical performance of a-Si thin film transistors ($\mu \approx 1 \frac{cm^2}{Vs}, \frac{On}{Off} \approx 10^6$).

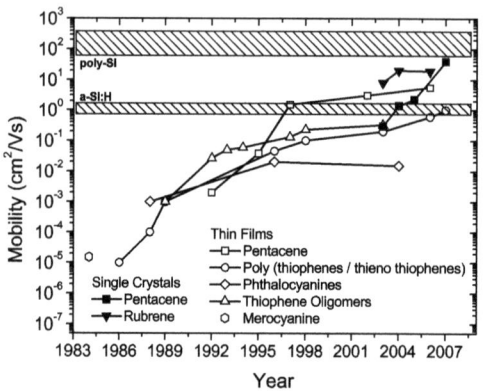

Figure 1.1: Development in OFET p-type mobility between 1984 and 2007. The various p-type materials are grouped together in families of similar molecular core part. The shaded bars represent the a-Si:H and Poly-Si mobility range [11].

The performance of organic single crystal devices is in fact approaching values of inorganic polycrystalline TFTs ($\mu \approx 50 - 400 \frac{cm^2}{Vs}, \frac{On}{Off} \geq 10^6$) [11]. The rapid improvement in organic transistor performance is illustrated in Figure 1.1 for the case of p-type OFET charge carrier mobility, summarizing the development of thin film polymer, small molecule and single crystal transistor devices from the year 1984 to 2007. The references for the illustrated mobility values of the selected materials are listed in Tables C.1 and C.3. A continuous increase in charge carrier mobility of over six orders of magnitude during the course of the last two decades is demonstrated by this graph. It is widely accepted, that this increase is due to optimized semiconductor solid morphology, as the result of improved process parameters as well as an adequate choice in smooth (rms ≈ 1nm) functionalized or non functionalized dielectrics [12–16]. While the semiconductor morphology is important, due to a required strong $\pi - \pi$ overlap in the charge carrier transport direction, additional factors for enhanced OFET performance are improved semiconductor purity[1] as well as the use of different or newly designed organic semiconductors. Furthermore, the transistor performance is influenced by charge carrier injection properties, which depend on the OFET configuration as well as the choice in source-drain contact metalization.

While p-type OFET charge carrier transport has been intensely studied since the first OFET implementation, only few reports about n-type organic transistors have been published in comparison. Usually, especially designed semiconductors are utilized for n-type OFET applications [17, 18], mostly comprising strong electron withdrawing groups, such as -F or -CN, since it was generalized until recently, that the unipolar charge carrier transport properties of organic semiconductors are an intrinsic [19, 20] material attribute. With regard to complementary metal oxide semiconductor (CMOS) technology, where p- and n-type transistors with balanced charge carrier transport properties are required on a single substrate, complicated and costly approaches for the realization of organic CMOS technology have therefore been implemented, in order to deposit different semiconductors on a single substrate. The complicated approaches are required, based on the intolerance of most organic semiconductors to UV radiation, water as well as oxygen, and therefore their incompatibility to classical lithography without significant degradation of the charge carrier transport properties. The most frequently implemented technique for the realization of CMOS devices is the deposition of spatially separated semiconductor materials by physical vapor deposition (PVD) [21–24]. This structured pre-

---

[1] A improved semiconductor purity results into a reduced charge carrier trap concentration (see chapter 2.1).

cipitation of the respective organic semiconductors is made possible by the use of shadow masks. An alternative approach could possibly be the use of inkjet printing [25–27].

However, for OFET applications it was recently demonstrated, that the field effect charge carrier transport properties of organic semiconductors can be influenced by electronic states, such as hydroxyl groups, at the dielectric / semiconductor interface [28, 29]. This is an aspect, which has been underestimated in the existing OFET development, due to the assumption, that organic semiconductors are not able to form dangling bonds, which are considered to be the main cause for interface states in inorganic semiconductors. As a result, it could be demonstrated, that pentacene, which was considered a material with unipolar p-type transport properties [20], exhibits balanced field effect charge carrier transport properties for both electrons and holes [28], if deposited onto an appropriate dielectric. In fact, Chua et al. [29] were able to demonstrate ambipolar transport in a variety of polymer semiconductors, deposited on different gate dielectrics. This indicates, that organic semiconductors, in dependence of their electron affinity, are actually intrinsically ambipolar [19] instead of unipolar n- or p-type. The demonstrated importance of dielectric interface states for OFET charge carrier transport properties leads to the question, of how dielectric interface engineering can be utilized, to modify ambipolar transistor charge carrier transport for organic CMOS applications.

This thesis deals with the application of two different approaches in dielectric interface engineering to address the challenge of realizing organic CMOS technology, using only a single organic semiconductor and in fact an identical device cross section. These techniques allow for the removal and introduction of charge carrier traps at the dielectric interface. In addition, utilizing the electret properties of a polymeric gate insulator, the influence of a forming step on the OFET charge carrier transport properties is investigated. The content of the thesis is structured as follows:

The theoretical and experimental basics for the understanding of the conducted investigations are described in *chapters 2 and 3*.

The possibility to realize unipolar p- and n-type pentacene OFETs, using an adequate source-drain metalization, is treated in *chapter 4*. Further, the influence of several polymeric gate dielectrics, with regard to unipolar transistor charge carrier transport is investigated. The insulators were chosen, due to a varying amount of oxygen containing polar groups in their monomeric unit.

Two approaches used to modify the dielectric interface for changes in the OFET charge carrier transport are dealt with in detail in *chapter 5*. In a first approach, the influence of Ca trace modified SiO_2 insulators [28] on the n-type pentacene OFET charge carrier transport properties is investigated. For these experiments, electrical transistor parameters as well as photoelectron spectroscopy measurements are considered in dependence of the Ca layer thickness. In the second approach, the influence of a polymethylmethacrylat insulator, exposed to ultra violet (UV) radiation in ambient atmosphere, is investigated with respect to pentacene OFET charge carrier transport. For these investigations, to quantify UV induced changes at the polymer interface, a detailed interface analysis is conducted, using atomic force microscopy, water contact angle as well as photoelectron spectroscopy measurements, in addition to the electrical characterization of the completed OFET device structure. In the final section of *chapter 5*, the investigated approaches for dielectric interface engineering are examined with regard to their applicability to organic CMOS technology.

In *chapter 6*, the influence of positive charges, stored in a polymethylmethacrylat insulator on the pentacene OFET charge carrier transport, is analyzed. For these investigations, the electret properties of the polymeric insulator are exploited by charging the dielectric, using a thermal forming step.

Chapter 2

Theoretical fundamentals

In the following chapter, the theoretical fundamentals for the understanding of this work are elaborated. However, no claim for an exhaustive treatment of the respective topics is made. For further information, the inclined reader is referred to more detailed literature [30–33].

2.1 Organic semiconductors

Definition

Carbon based molecules, that allow for the injection and transport of charge carriers, due to a conjugated π-electron system, and feature optical gaps, spanning the energy spectrum from the near ultra violet to the near infrared energy range, are referred to as organic semiconductors in the following.

Basic properties

Organic semiconductors are classified into two material classes, which are small molecules and polymers. While the electronic properties of both types of materials are more or less the same, they differ with respect to their processability. Thin films or single crystals of small molecules are typically grown by physical vapor deposition. Polymers, however, cannot be thermally evaporated, due to a deterioration of their molecular structure during the process. This is the result of their high molecular weight. However, the design of most polymers allows this type of semiconductor to be processed from solution. Technologically, this is achieved by engineering monomer unit side chains, that impede the crystallization of the polymer in solution.

For organic semiconductors, the σ-bonds of sp^2-hybridized C-atoms form the backbone of the respective molecules, due to their covalent bonding nature. The additional overlap in p_z orbitals for neighboring C-atoms leads to a further bonding type, the π-bond, which is responsible for the delocalized electron system along the conjugation path of the respective molecules and therefore their semiconducting properties. In a perfect system, this would lead to a delocalization along the entire conjugation length of the molecule. However, in real systems, the electron delocalization is limited by impurities, imperfections and twists in the molecular structure. Due to the limited spatial dimensions of small molecules and the possibility to purify the organic semiconductor, for example by gradient sublimation, the π-electron system is generally assumed to be delocalized along the entire conjugation length. For polymers with their much larger spatial dimensions, however, the delocalization is typically limited to a few monomer units.

The split of the π-bonds in their binding (π) and anti-binding (π*) states, for molecules in their gas phase, defines the highest occupied molecular orbital (HOMO) and the lowest unoccupied molecular orbital(LUMO). For the case of pentacene, these frontier orbitals, as experimentally obtained by measuring the ionisation potential I and the electron affinity χ [31], are illustrated in Figure 2.1 1).

By considering solids, one of the most important differences to inorganic semiconductors is the preservation of the molecular properties in organic compounds such as chemical identity, vibrionic oscillations as well as the conservation of the molecular electronic structure. This is due to the weakness of the van der Waals interaction forces [2], which are responsible for the cohesion of the molecular solids. For organic solids, these forces are primarily due to easily polarizable π-orbitals with fluctuating electron densities and therefore time dependent dipole moments [31]. With regard to charge carrier transport states, the weak intermolecular interaction forces for organic semiconductors are the cause for significant differences, when compared to their inorganic counterparts. For inorganic semiconductors, the electronic wave functions are delocalized throughout the entire semiconductor, resulting in energetically wide quasi continuous valence and conduction bands of up to 8eV. This energy range allows for band type transport. For solids of molecular semiconductors, however, the delocalization of the electronic wave function is weak, resulting in a split of the HOMO / LUMO levels into energetically narrow bands of several 100meV. Therefore, in accordance with the frontier orbitals of organic molecules in their gas phase, solid state transport states of molecular semiconductors are designated as HOMO / LUMO levels in the following.

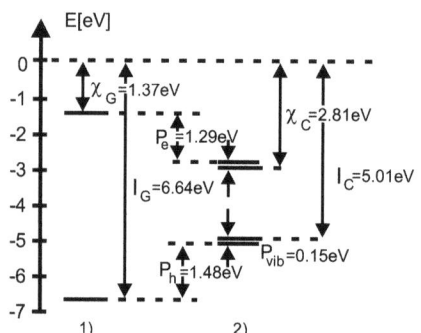

Figure 2.1: Pentacene energy diagram with respect to the electron affinity and the ionisation potential. The values are depicted for the gas phase 1) (χ_G, I_G), as well as the single crystal form 2) (χ_C, I_C) of pentacene. [31]

In solids the energetic position of the HOMO / LUMO levels changes due to electronic or vibronic polarization effects, as indicated in Figure 2.1 2). Here, the respective energy states are illustrated for pentacene in its monocrystalline form [31]. The electronic polarization of these states by a charge carrier, either injected or thermally generated within the molecular solid, is > 1eV. This is indicated by P_e as well as P_h for electrons and holes. This results in an energy gap of 2.2eV between the HOMO / LUMO transport levels for the case of pentacene. In general, solids of organic semiconductors exhibit energy gaps between 1.5eV and 3eV. Since the charge transfer time from one molecule to the next is by a factor 100 larger than the time required

---

[2] Van der Waals forces are defined by weak electrical dipole forces between neutral molecules that exhibit fully occupied molecular orbitals.

2.1 Organic semiconductors

for the charge carrier to polarize its molecular environment, the polarization is bound to the charge carrier during its transport through the molecular solid. This is called a polaron. In addition to the electronic polarization of the energy states, the molecular polarization[3] leads to a further reduction in the optical bandgap of the organic molecular solid by ≈ 0.3eV [31].

The susceptibility of transport states of organic solids to polarization effects makes it necessary, to differentiate between organic molecular crystals and organic molecular disordered solids, when considering charge transport properties.

Experiments on **organic molecular crystals**, conducted by Warta et. al [34], have demonstrated an increase in charge carrier mobility for a decrease in sample temperature. This experimental result was ascribed to a reduction in phonon scattering and implies band like charge carrier transport [32,34] for organic molecular crystals. Indeed, Cheng et. al [35] were able to confirm the possibility of band like transport for organic molecular crystals by using Hartree-Fock-INDO calculations, to estimate the bandwidth of several polyacene crystals. For the case of pentacene, a total HOMO / LUMO bandwidth of ≈ 700meV was determined. The energetically narrow bands develop in consequence of a π-π orbital overlap for adjacent molecules, due to the high translational symmetry of molecular crystals. However, it was also substantiated, that the band transport for organic molecular crystals is limited to temperatures below $T \approx 150$ K [35], due to electron phonon coupling. For $T > 150$ K, the mean free path length of charge carriers is in the order of ≈ 5Å, which is in the order of the dimensions for a molecular unit cell[4]. This reduction in mean free path length marks the transition from band like to hopping transport. The influence of the structural order on the charge carrier mobility is estimated, by considering the effective mass of charge carriers at the band edge m^*. By taking the Hueckel-Theory into account [37], m^* can be approximated by:

$$m^* = \frac{\hbar^2}{2a_0^2 \beta} \quad (2.1)$$

Here, a_0 represents the distance between two molecules, and the parameter β describes the transfer-integral energy. A good overlap in π orbitals, due to structural order, results in an increase in β and therefore in a reduction of the charge carrier effective mass. By considering the following Drude approximation for the charge carrier mobility (equation 2.2), a reduction in m^* and therefore an enhanced structural order should lead to an increase in charge carrier mobility.

$$\mu = \frac{e}{m^*} * \frac{\tau}{2} \quad (2.2)$$

Here e represents the elemental charge and τ represents the mean free time.

The charge carrier mobility in **disordered molecular solids** is rising with increasing temperatures [38], implying a hopping transport even at $T < 150$ K. This is the result of a missing translational symmetry as well as weak intermolecular binding energies, which impede the formation of transport bands. For such compounds, the transport gap between the HOMO / LUMO transport levels is not

---

[3] Polarization due to interatomic oscillation.
[4] The pentacene molecule has spatial dimensions of 7.9Å, 6.06Å and 16.01Å in the respective a, b and c directions. Two molecules are arranged per unit cell [36].

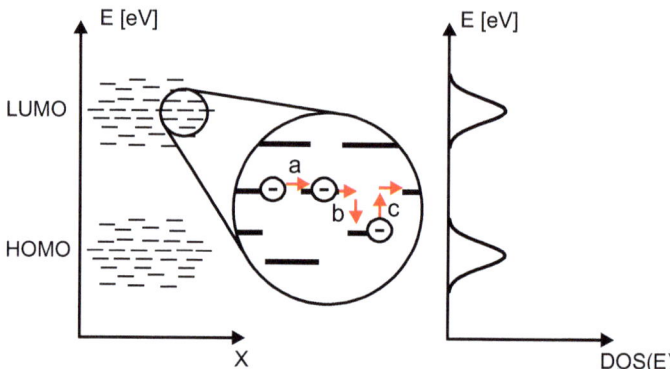

Figure 2.2: Transport states in a disordered molecular solid. Three different hopping transitions a), b) and c) are illustrated by the inset. Only for transition c), an activation energy is required prior to the tunneling process.

only defined by intermolecular interactions, but may vary significantly, due to the disorder of the solid. The disorder and hence the locally different polarization environments result in a distribution of the respective HOMO / LUMO levels, that is described in the following by a Gaussian Density Of States (DOS) [39] distribution, as depicted by the left side of Figure 2.2. The DOS is quantified by equation 2.3:

$$DOS(E) = \frac{N}{\sqrt{2\pi}\,\sigma}\,exp\left(-\frac{(E)^2}{2\sigma^2}\right) \qquad (2.3)$$

Here N represents the spatial density of states, and the energy E has to be considered with respect to the energetic center of the DOS. The typical standard deviation / width for organic disordered solids is in the order of $\sigma \approx 100\,meV$ [40].

Hopping transport

The charge carrier transport in organic molecular crystals, where the localization of a charge carrier on a molecule prevails, as well as disordered molecular solids can be described by a hopping transport. This is the direct result of an energetical disorder of the HOMO / LUMO transport levels. The assumed Gaussian density of states distribution for disordered molecular solids is exemplified by the illustration of Figure 2.2. In principle, the hopping transport can be described as a tunneling process. However, tunneling is an isoenergetic event, which needs to be considered, when discussing the actual transport. In the following, the hopping process is covered, using a model first introduced by Bässler [39]. The hopping rate between two transport states can be quantified by the following Miller-Abraham type equation [41]:

$$\nu_{ij} = \nu_0 exp(-2\gamma \Delta R_{ij}) \begin{cases} exp(-\frac{E_j - E_i}{kT}); & E_j > E_i \\ 1; & E_j \leq E_i \end{cases} \qquad (2.4)$$

Here E_i and E_j represent the respective initial and target energies for the hopping process within the DOS . The norm distance between these states is defined by ΔR_{ij}, and the constant γ symbolizes the exponential decay of the respective wave functions. The Boltzmann constant is given by k and the temperature by the parameter T. The equation is a product of the attempt to escape frequency v_0, an exponential factor describing the tunneling probability, as well as a Boltzmann factor for hops upwards in energy ($E_j > E_i$). Hops down in energy ($E_j < E_i$) are not impeded by the energy matching condition of the tunneling. The surplus energy is dissipated after the tunneling process.

Illustrated by the inset of Figure 2.2 are three possible hopping occurrences a, b and c, in order to visualize the transport. For transitions a) (isoenergetic tunneling) and b) (energy dissipation after the isoenergetic tunneling), no thermal activation is necessary prior to the tunneling process. Such a thermal activation, however, is needed for transition c), where a hop upwards in energy is required for the transport in X direction.

As discussed above, the charge carrier transport in molecular crystals as well as disordered molecular solids is dominated by a hopping transport for temperatures > 150 K. However, a good structural order of the solid, as it is found in molecular crystals, is still beneficial for the charge carrier transport. As previously discussed, an overlap in π-orbitals leads to an enhancement in transferintegral energy and therefore into an increase in charge carrier mobility.

Charge carrier traps

In addition to what has been stated above, possible energetic states in the band gap above or below the HOMO / LUMO transport levels function as charge carrier trap states. Furthermore, low tail states in the HOMO or LUMO DOS may act in a similar fashion. Once a charge carrier is localized within these states, it no longer contributes to the charge transport, unless it is thermally activated to escape its trap. For the case of electrons, trap states are considered energetically deep for energy values $E_T \geq E_F + kT$ [42]. For the case of holes, the inequality $E_T \leq E_F - kT$ holds. The trap level energy E_T as well as the Fermi level E_F are considered with respect to the vacuum level. The distribution of the trap states can range from monoenergetic states to a random distribution.

The origin of semiconductor trap states is the result of a morphological or chemical nature. Morphological traps are usually due to grain boundaries or a disorder of the molecular solid. Chemical traps, on the other hand are due to neutral doping or impurities of the organic semiconductor as well as defects in the monomeric units or chain irregularities. Furthermore, functional groups containing oxygen can significantly influence the charge carrier transport by the localization of negative charge, as demonstrated for the case of keto groups in polyfluorene-type conjugated polymers by Kadashchuk et al. [43].

For the special case of field effect transistors, where the charge carrier transport occurs directly at the dielectric / semiconductor interface, electronic interface states may also influence the charge carrier transport [44]. This is exemplified by the illustration of Figure 2.3, where the electron trapping mechanism of hydroxyl groups, available at a SiO_2 / organic semiconductor interface is depicted, as proposed by Chua et al. [29]. It is suggested, that the trapping of an electron occurs, due to the dissociation of a hydrogen atom, leading to the formation of a negatively charged silicon ion at the dielectric interface. While hydroxyl groups function as electron traps, they do not seem to influence the charge carrier transport of holes, as will be dealt with in chapter 4.2. This indicates, that complementary charge carrier traps can exist independently of one another [45].

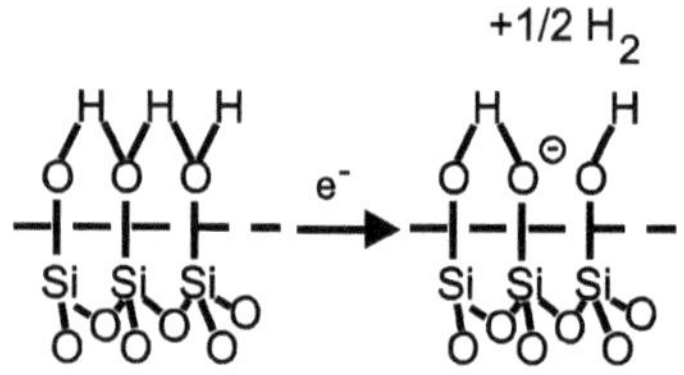

Figure 2.3: Hydroxyl group electron trapping mechanism suggest by Chua et al. [29].

2.2 Transistors

Since its first realization by Bardeen, Shockley and Brattain in 1951 [46], the transistor has become one of the key elements in modern day electronics. Its widespread implementation is at last the result of the capability, to modulate the electric current by an electric quantity in a solid state device. This is a concept, which was first introduced by Julius Edgar Lilienfeld in 1928 [47], who described the control of the resistor conductivity by a pure solid state effect. Even though many approaches for the transistor realization have been developed over the years, they can mainly be subdivided into two groups of bipolar and field effect transistors (FETs):

- Bipolar transistors are realized, using either a npn- or pnp-junction sequence. The different areas of the device are designated as collector, base and emitter electrode. To operate the transistor, one of the junctions is forward biased, where as the other is biased in reverse. Using a small control current over the base electrode, a significant current between the collector and emitter electrodes is enabled.

- As for the bipolar transistor, the field effect transistor incorporates an electrode (Gate), by which the current between two further electrodes (Source and Drain) can be modulated. However, for this type of transistor, the control electrode is isolated from the transistor channel by a gate dielectric, as depicted in Figure 2.4(a). The conductivity in the transistor channel is altered by applying a voltage to the gate electrode, resulting in a change in charge carrier density at the dielectric / semiconductor interface as a consequence of the field effect. In contrast to the bipolar transistors, the field effect transistor is voltage controlled.

However, for organic transistor applications, the realization of bipolar transistors is difficult, and has therefore impeded its widespread application. This is due to the circumstance, that stable charge transfer doping is not well established as a result of high diffusion of the respective dopants, making the realization of the necessary npn- or pnp-junction sequence not easily feasible. For the work at hand, only organic field effect transistors are discussed.

2.2 Transistors

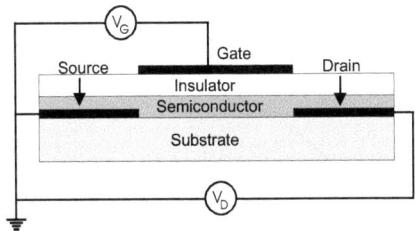
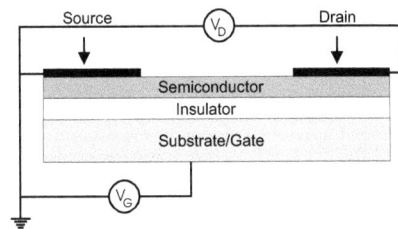

(a) OFET top gate and bottom source-drain contact configuration.

(b) OFET bottom gate and top source-drain contact configuration.

Figure 2.4: Standard OFET designs in top / bottom source-drain contact configuration. The top / bottom gate architecture indicated in Figures 2.4(a) and 2.4(b), can be applied to both of the source-drain contact configurations.

2.2.1 Organic field effect transistors

The organic field effect transistor is a type of thin film transistor (TFT), which differs from the standard FET structure by its intrinsic semiconductor layer as well as by its processability. Inorganic TFTs are widespread in todays electronic applications, since its structure is compatible with common thin film techniques, such as physical or chemical vapor deposition, as well as solution based processes, such as dip or spin coating. The TFT is therefore not limited to a specific substrate, which represents one of its main advantages. Organic field effect transistors represent an interesting extension to the TFT family. This is due to the promise of cheap role to role processability, or the implementation of transistors on plastic substrates for applications, such as flexible displays or RFID tags.

Illustrated in Figures 2.4(a) and 2.4(b) are two typical standard OFET designs. Depicted in Figure 2.4(a) is a top gate configuration comprising, bottom source-drain contacts, while in Figure 2.4(b) a bottom gate configuration with top source-drain contacts is illustrated. Both top and bottom gate configurations are also commonly implemented with a respective top or bottom source-drain architecture (not shown). The resulting typical four transistor structure combinations can be chosen in dependence of the application or material requirements.

The working principle of an OFET is based on the field effect, where mobile charge carriers, either thermally generated[5] or injected from the source-drain electrodes, are accumulated at the insulator / semiconductor interface to compensate the electric field applied by the gate electrode. As a result, the total amount of accumulated charge n and therefore the conductivity σ in the transistor channel can be controlled by the applied gate potential. The conductivity is defined by $\sigma = \sum_i \mu_i n_i e$, where μ represents the charge carrier mobility and e defines the elemental charge. A current through the transistor is obtained, by applying a lateral electric field between the source-drain electrodes as long as the conductivity of the semiconductor layer is sufficiently high. The actual current transport occurs within the so called transistor channel. This channel is spatially limited to the charge carrier

---

[5] For the following discussion thermal generation of charge carriers is neglected, since the time constant for charge carrier injection is significantly shorter [48].

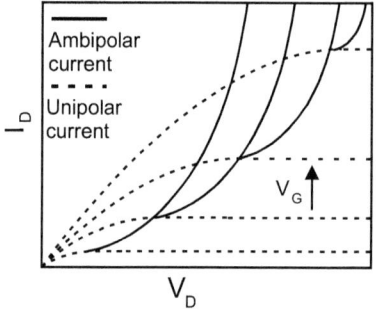 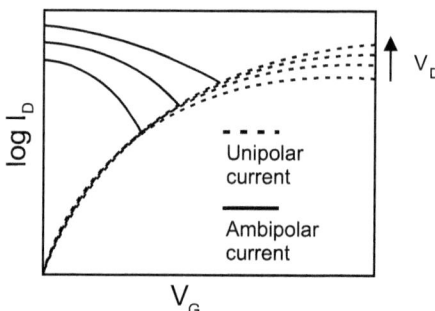

(a) I_D in dependence of V_D describing the transistor output characteristic.

(b) I_D in dependence of V_G representing the transistor transfer characteristic.

Figure 2.5: Schematic illustration of typical transistor current voltage characteristics. For $V_D \leq V_G$ the transistor is always operated in the unipolar range (dotted lines). For $V_D > V_G$ uni- or ambipolar (dotted lines) operation is possible. OFETs that allow for the injection and transport of electrons as well as holes exhibit ambipolar operation in that voltage range. Otherwise, the transistor characteristic saturates and remains unipolar.

accumulation zone at the insulator / semiconductor interface, typically extending over the first few monolayers of the organic semiconductor. This spatial limitation of the transport channel indicates the importance of the dielectric / semiconductor interface for the charge carrier transport in transistors [44].

Typically, a transistor is characterized by two interpretations of the current voltage characteristic, which are schematically illustrated in Figures 2.5(a) and 2.5(b). The dotted lines represent the unipolar device characteristic, while for the case of an ambipolar transistor behavior ($V_D \geq V_G$), only the solid extensions of the characteristics have to be taken into account. Ambipolar transistor behavior occurs, if the applied drain voltage V_D exceeds the applied gate voltage V_G, the source-drain contacts allow for the injection and the semiconductor permits the transport of complementary charge carriers. Therefore, in contrast to a unipolar transistor, where the drain current is driven by only one type of charge carrier, two complementary charge carrier types contribute to the drain current of an ambipolar field effect transistor. A more descriptive discussion of the ambipolar charge carrier transport is elaborated below. The transistor output characteristic is shown in Figure 2.5(a), which describes the drain current I_D in dependence of the drain voltage. This figure is usually considered as an indicator for a non ohmic contact resistance between the source-drain metalization and the organic semiconductor. For $V_D \ll V_G$, the characteristic exhibits a linear / s-shaped behavior for respective low / high injection barriers. Here, the output characteristic is depicted for the case of ohmic contacts. Illustrated in Figure 2.5(b) is the transistor transfer characteristic, which describes the drain current in dependence of the gate voltage. This characteristic is usually considered, in order to extract the device charge carrier mobility as well as its threshold voltage by the use of the Shockley transistor model [49].

2.2 Transistors

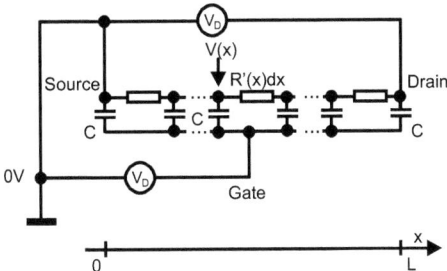

Figure 2.6: Resistor capacitor equivalent circuit for the extended Schockley transistor model [33].

In the following, an extension of the Shockley model is discussed, as published by Schmechel et al. [33], in order to describe the uni- and ambipolar transistor currents. The model is based on the gradual channel approximation, and was derived under the assumption of ohmic source-drain contacts as well as an infinite recombination probability between electrons and defect electrons (holes) in the transistor channel. Charge carrier density dependent mobilities are not considered.

The approach to calculate the transistor drain current is based on a simple resistor capacitor equivalent circuit, as depicted in Figure 2.6. The drain current is obtained by applying Ohm's law (2.5).

$$I_D = \frac{V(x)}{R(x)} \tag{2.5}$$

Here, $V(x)$ represents the voltage drop along the insulator surface between the source electrode and the position x, as indicated by the equivalent circuit. The differential resistance in the transistor channel is given by the parameter $R'(x)$, which allows for the derivation of the partial channel resistance $R(x)$ between the position x and the source electrode at $x = 0$. The resulting partial channel resistance is described by equation 2.6.

$$R(x) = \int_0^x \frac{dx}{We[\mu_n n(x) + \mu_p p(x)]} \tag{2.6}$$

Here, the respective electron and hole charge carrier mobilities are defined by μ_n and μ_p. W represents the transistor channel width. The area charge carrier density per unit area for electrons and holes is described by $n(x)$ and $p(x)$ respectively. However, as long as these parameters are not known, equation 2.6 is undefined.

In the following, $R(x)$ is determined by using the net surface charge $Q(x)$ per unit area, as given by equation 2.7:

$$Q(x) = e[n(x) - p(x)] = C[V(x) - V_G] \tag{2.7}$$

C represents the capacitance per unit area, and V_G describes the applied transistor gate voltage. Taking into account, that $V(x)$ is defined by the voltage divider, depicted in Figure 2.6, the voltage drop can be described by equation 2.8:

$$V(x) = V(x_0)\frac{R(x)}{R(x_0)} \tag{2.8}$$

$R(x_0)$ symbolizes the device channel resistance, integrated up to a random position x_0 in the transistor channel. By substituting equations 2.8 and 2.6 into equation 2.7, the following differential equation for a local variation of the channel's net charge is obtained by the differentiation of equation 2.7:

$$\frac{dQ}{dx} = \frac{CV(x_0)}{R(x_0)} \cdot \frac{1}{We[\mu_p p(x) + \mu_n n(x)]} \quad (2.9)$$

Under the approximation of an infinite recombination probability[6], equation 2.9 can be solved by assuming $Q(x) = ep(x)$ for holes, or $Q(x) = -en(x)$ for electrons.

If the transistor is operated in the unipolar mode($|V_D| \leq |V_G|$), only one charge carrier type is accumulated in the transistor channel. Therefore, by taking into account the boundary conditions $x_0 = L$, $V(x_0) = V_D$ as well as $Q_0 = CV_G$ at the source electrode, the integration of equation 2.9 yields the net charge per surface area:

$$Q(x) = \begin{cases} -\sqrt{C^2 V_G^2 - \frac{2C}{\mu_n W} \cdot \frac{V_D}{R(L)} \cdot x} & \text{for } V_G > 0 \text{ (electrons)} \\ \sqrt{C^2 V_G^2 + \frac{2C}{\mu_p W} \cdot \frac{V_D}{R(L)} \cdot x} & \text{for } V_G < 0 \text{ (holes)} \end{cases} \quad (2.10)$$

By using the boundary condition $Q(L) = C(V_D - V_G)$, $R(L)$ can be determined:

$$R(L) = \left| \frac{L}{\mu_{n/p} \cdot WC(\frac{1}{2}V_D - V_G)} \right| \quad (2.11)$$

As already indicated above, for $|V_D| > |V_G|$, the transistor is operated in the ambipolar current-voltage range. This means, that both electrons and holes contribute to the transistor drain current, as long as the injection barriers for both charge carrier types are sufficiently low and the semiconductor allows for ambipolar behavior. Under the assumption of an infinite recombination probability between electrons and holes, the transistor channel is separated into a unipolar electron and hole transporting section, as indicated in Figure 2.7, for a transistor operated in the ambipolar range. The position x_0 in the transistor channel forms the intersection between the two unipolar transport ranges, where the recombination of the complementary charge carriers occurs. At this position, the accumulated net charge is zero and consequently the potential is given by $V(x_0) = V_G$. For the ambipolar range, the total channel resistance is composed of two components, as described by equation 2.12.

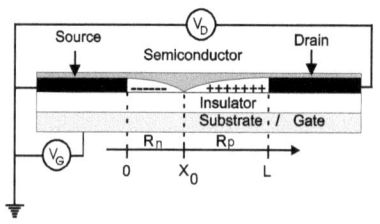

Figure 2.7: Charge carriers contributing to the ambipolar current for $|V_D| > |V_G|$

$$R(L) = R_n + R_p \quad (2.12)$$

R_n represents the channel resistance from the source electrode to the position x_0 and R_p defines the channel resistance between x_0 and the drain electrode. By considering that $Q(x_0) = 0$, and the delineations above, the

---

[6] This results into a concentration of zero for electron-hole pairs in the transistor channel.

2.2 Transistors

resistances R_n and R_p can be quantified as:

$$R(n) = \frac{2x_0}{\mu_n WC |V_G|}, \qquad R(p) = \frac{2(L - x_0)}{\mu_p WC |V_D - V_G|} \tag{2.13}$$

The mobilities of the complementary charge carrier types are represented as μ_n and μ_p. Due to a constant current in the transistor channel, the position of the recombination zone (x_0) can be derived by taking into account $\frac{|V_G|}{R_n} = \frac{|V_D - V_G|}{R_p}$ as:

$$x_0 = \frac{LV_G^2}{V_G^2 + \frac{\mu_2}{\mu_1}(V_D - V_G)^2} \tag{2.14}$$

Finally, the transistor drain current can be derived for its uni- and ambipolar range, by subsequently substituting equations 2.11 and 2.13 / 2.14 into equation 2.5. For the previous calculation, a fully depleted transistor channel at zero gate bias was assumed as well as a contribution of all charge carriers accumulated in the transistor channel, due to an applied gate field, to the charge carrier transport. These assumptions ignore trap states in the transistor channel or at the dielectric / semiconductor interface, which may localize charges otherwise available for the current transport. Furthermore, it was ignored, that mobile charge carriers are possibly already available in the transistor channel at zero gate bias. To account for these effects, which influence the effective gate voltage, a threshold voltage $V_{th,n}$ and $V_{th,p}$ for electrons and holes is introduced to the drain current equations.

In the following, the drain current equations are given for the uni- and ambipolar range in the electron accumulation mode ($V_G > 0$). The equations 2.15 - 2.17 have been derived under the condition of $V_{th,n} > 0 > V_{th,p}$.

(1) Unipolar range $V_D \leq (V_G - V_{th,n})$

$$|I_D| = \frac{WC}{L}\mu_n[(V_G - V_{th,n}) - \frac{1}{2}V_D]V_D\Theta(V_G - V_{th,n}) \tag{2.15}$$

(2) Saturation range $V_D \geq (V_G - V_{th,n})$ but $V_D \leq (V_G - V_{th,p})$

$$|I_D| = \frac{WC}{2L}\mu_n(V_G - V_{th,n})^2\Theta(V_G - V_{th,n}) \tag{2.16}$$

(3) Ambipolar range $V_D \geq (V_G - V_{th,p})$

$$|I_D| = \frac{WC}{2L}[\mu_n(V_G - V_{th,n})^2\Theta(V_G - V_{th,n}) \\ + \mu_p(V_D - (V_G - V_{th,p}))^2] \tag{2.17}$$

$\Theta(x)$ stands for the Heaviside step function, which is defined as $\Theta(x) = 0$ for $x \leq 0$ as well as $\Theta(x) = 1$ for $x \geq 0$. The described model will be utilized in the following sections, to derive transistor parameters such as the mobility or the threshold voltages.

2.2.2 Influence of the dielectric interface on charge carrier transport in organic field effect transistors

The dielectric interface surface of an OFET influences the current transport mainly by (1) influencing the morphology of the organic semiconductor, (2) the dielectric properties of the insulator and (3)electronic states at the dielectric interface.

Aspect 1) The influence of the dielectric interface on the semiconductor morphology, with respect to grain size and molecular orientation, is probably the most frequently investigated issue. The most common approaches are, to minimize the insulator surface roughness, and to control the orientation of the molecules by the use of self assembled monolayers (SAMs) [12–16, 50]. This is important, as outlined above, due to the required strong π-π overlap in the transport direction as well as the non-isotropic charge carrier transport in organic semiconductors [31]. Recent studies [24, 51] have demonstrated, that SAMs can even be used as thin (d=2.5nm) dielectrics, in order to realize low power OFET and circuitry applications.

Aspect 2) The effect of dielectric properties, such as the dielectric constant, on the charge carrier transport has so far been less often investigated. However, there is experimental evidence for a decrease in mobility with an increase in the dielectric permitivity [50, 52, 53]. This effect is most probably due to a broadening of the DOS, as the result of polarization effects, and therefore a reduced charge carrier mobility

Aspect 3) The effect of electronic states at the dielectric interface has been underestimated so far for organic field effect transistors and is therefore less frequently investigated. This is due to the lack in dangling bonds for organic semiconductors, which are the main cause for electronic interface states in their inorganic counterparts. However, it could be demonstrated, that pentacene, which is well known for its unipolar hole transporting properties, also conducts electrons [28, 55] with a mobility comparable to that of holes. The inhibited electron field effect mobility of the well researched material could be linked to electron traps at the dielectric / semiconductor interface [28], in particular on SiO_2 surfaces, which have been the insulator of choice for most investigations. The importance of a trap free dielectric interface for the OFET charge carrier transport is further substantiated by the following metal insulator semiconductor (MIS) diode experiment [54].

The investigated device has a cross section, consisting of a p^{++} – Si / SiO_2 /pentacene / Ca layer stack, analogue to the typical OFET bottom gate / top source-drain contact device structure. Illustrated in Figure 2.8(a) is the differential MIS diode capacitance measurement of this experiment, using a measurement frequency of 100Hz. C_{Diel} represents the capacitance of the 200nm SiO_2 dielectric, while C_{Tot} stands for the total device capacitance including the organic semiconductor. The MIS diode shows no significant dependence of its capacitance on the applied voltage (V_G). For reverse bias, this is due to a large energy barrier at the Ca / pentacene interface, suppressing the hole injection. For forward bias, the negligible response in the capacitance voltage characteristic is suggested to be either due to an insufficient electron injection or to the circumstance, that injected electrons cannot be transported through the pentacene bulk at the applied measurement frequency, or possibly to electron traps at the dielectric interface.

2.2 Transistors

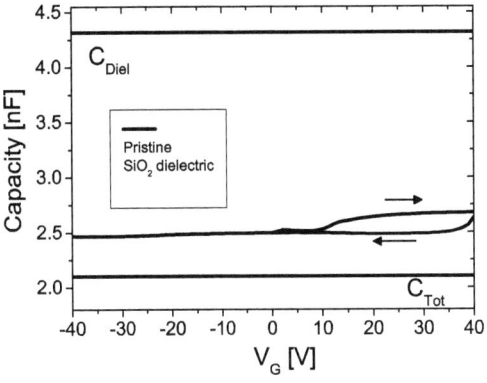
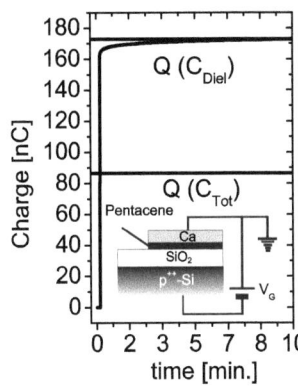

(a) Impedance measurement using a frequency of 100Hz, for devices incorporating a pristine SiO$_2$ dielectric

(b) MIS diode charging experiment at an applied DC bias of $V_G = 40V$. Inset: MIS diode structure

Figure 2.8: Impedance and DC measurements on MIS diodes consisting of a p^{++}-Si / insulator / pentacene / Ca layer stack [54]

By taking Figure 2.8(b) into account, which illustrates a DC charging of the diode at an applied voltage of V_G=40V, an accumulated charge of 172nC is determined. This amount of charge corresponds to a device capacitance of C=4.3nF[7], which is close to the value obtained for C_{Diel}. As discussed by Ahles et al. [28], this proves, that electrons can be injected into the organic semiconductor and transported through the bulk of the pentacene layer. However, once accumulated at the dielectric interface, the negative charge carriers cannot follow the applied AC electric field of the measurement, which is expressed by the lack of change in the differential device capacitance, illustrated in Figure 2.8(a).

This result strongly suggests, that charge carrier traps are available at the interface of SiO$_2$ insulators, impeding the electron current transport in pentacene OFETs incorperating such a dielectric. The drawn conclusions were further supported by considering the impedance measurement of a MIS diode with Ca modified SiO$_2$ dielectric (not shown). For such a device, the applied forward bias measurement yielded a differential capacitance value close to that of the device dielectric. This indicates, that the suggested interfacial electron traps are influenced by the Ca modification. As discussed above, these traps were later identified by Chua et al. as hydroxyl groups.

The effects of electronic states at the dielectric interface, with respect to the OFET charge carrier transport, will be the focus of this dissertation. In the following chapters, several techniques, to modify such trap states at the dielectric interface for the control of the OFET charge carrier transport, will be demonstrated. Furthermore, the introduced dielectric interface engineering approaches will be implemented, to realize OFETs with complementary charge carrier transport properties, however, with an identical device structure.

---

[7] A dielectric constant of $\epsilon = 3.9$ is considered for SiO$_2$ [50]

2.3 Electrets

The charging of an electret dielectric will be used, to influence the OFET charge carrier transport properties, as elaborated in chapter 6.

An electret is a piece of dielectric material, exhibiting a *quasi – permanent electrical charge*. The term "quasi-permanent" means, that the time constant characteristic for the decay of the charge carrier concentration in the electret is much longer than the time period over which studies are performed [30].

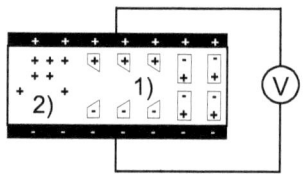

Figure 2.9: Electret charging by contacting electrode. 1) Dipole alignment and space charge separation. 2) Charge carrier injection due to high electrical field strenghts.

The type of charges which are stored in an electret are either space charges, the result of a true polarization of the electret, or a combination of the two. A multitude of approaches exist to charge electrets, such as electrical discharges, particle beams, contact electrification or by contacting electrodes. The following discussion will concentrate on electret charging by contacting electrodes, using a thermal charging method.

Thermal charging of an electret consists of the application of an electric forming field E_{form} to the dielectric, at an elevated forming temperature T_{form}. Subsequently, the forming temperature is cooled down, while the forming field is held at a constant value. During this process, two kind of charging phenomena can occur. 1) Internal polarization, due to dipole alignment or charge separation within the electret. 2) Charge injection through contacting electrodes. Both charging phenomena are schematically illustrated in Figure 2.9.

1. *Dipole alignment* under the influence of an electric field occurs at elevated forming temperatures (above the glas temperature for the case of a polymer insulator), where the molecules or molecular chains of the solid are sufficiently mobile. By cooling the electret, while the electric field strength is held constant, the dipoles are "frozen" in their aligned position, giving the electret a permanent polarization. It has been demonstrated by van Turnhout [56], that the electret polarization by dipole alignment is weakly forming-field and strongly temperature dependent.

    The polarization of an electret, due to *charge separation* of electret space charges, results from the temperature dependent conductivity $\sigma(T)$ of the electret, as indicated by equation 2.18:

    $$\sigma(T) = \sigma_0 exp(\frac{-E_C}{k_B T_{Form}}) \qquad (2.18)$$

    E_C represents the activation energy for conduction. The conductivity $\sigma(T)$ of the solid is increased with elevated temperatures, allowing for this type of polarization. The charge separation process exhibits the same forming field and temperature dependence as the dipole alignment process [30].

2. The *injection* and storage of excess charge in an electret can occur, if the applied electric field strength is sufficiently high for charge carriers to overcome the injection barrier between the

2.3 Electrets

injecting electrode and the electret. The charges are then localized in available trap states as discussed above. While the charge carrier injection is strongly forming-field dependent [57], elevated temperatures are still beneficial for this kind of electret charging process, due to an enhanced electret conductivity, as demonstrated by equation 2.18. Elevated temperatures allow for the excess charge to be transported into the bulk of the insulator. This allows the filling of energetically deep traps in the volume of the dielectric, which extends the discharge time of the electret at room temperature.

Chapter 3

Experimental Framework

In the present chapter, the required experimental framework for the investigations discussed in chapters 4 - 6 is delineated. At first the employed materials are presented, followed by a discussion of the applied thin film deposition techniques, the sample structure as well as specific techniques and conditioning methods, to characterize and alter the electrical device performance. In a final section of this chapter, the characterization techniques used to investigate the chemical composition or structure of certain sample layers are summarized.

3.1 Materials

Organic insulators

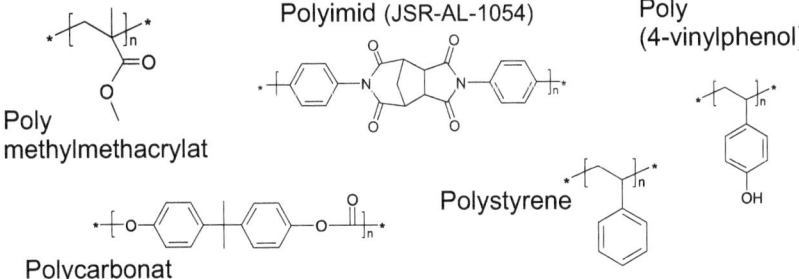

Figure 3.1: Chemical structure of the polymers used as dielectric materials. The abbreviations used for the individual polymers are summarized in Table 3.1.

The work at hand investigates, among other aspects, the influence of several polymeric gate dielectrics (*polymethylmethacrylat*, *polyimid*, *poly*(4 − *vinylphenol*), *polycarbonat* and *polystyrene*) on the charge carrier transport in pentacene OFETs. The chemical structure of these polymers is depicted in Figure 3.1. Furthermore, their respective dielectric constant and water contact angle values are listed in Table 3.1. The water contact angle values, determined as described in section 3.3, are compared to corresponding literature values. The determined values correspond well with literature

23

Polymer Name	Abbreviation	Dielectric Constant	Contact Angle [°] Determined	Contact Angle [°] Literature
Polystyrene	PS	2.5 [58]	97	102 [59]
Polycarbonat	PC	3.5 [58]	92	70 [60]
Poly-methylmethacrylat	PMMA	3.5 [50]	81	70 [61], 73 [62], 82 [63]
Poly(4-vinylphenol)	P4VP	4.5 [50]	76	-
Polyimid	PI	4 [64]	67	65 [65]

Table 3.1: Specifications of the utilized insulating polymers.

values, with the exception of PC. The observed discrepancy may be the result of different experimental conditions, as summarized in section 3.3. The difference in water contact angle for the respective polymers is suggested to be the result of varying amounts of oxygen containing polar groups, such as hydroxyl and keto groups, in the repeating chain of the material. The exact process parameters for the deposition of these polymers by the use of a spin coater, as well as their respective dilutions, are listed in Appendix D.

Pentacene

The small molecule organic semiconductor pentacene belongs to the family of the oilgoacenes and is the semiconductor of choice for the following experiments. This semiconductor has been subject to extensive investigations during the last 15 years, either in its crystalline form or as polycrystalline thin films. This intensive interest is mainly due to its exhibited high field-effect mobilities, as illustrated in Figures 1.1 and 4.1, as well as the promise for commercialization of products incorporating this type of semiconductor.

The pentacene molecule, as depicted in Figure 3.2(a), is a planar molecule[8] composed of five benzene rings ($C_{22}H_{14}$). In its bulk phase, pentacene has a triclinic structure [66], with two molecules arranged per unit cell [36]. The semiconductor HOMO / LUMO levels are illustrated in Figure 3.2(b) by the ionisation potential / electron affinity at respective energy values of $I_C \approx 5.01 eV$ and $\chi_C \approx 2.81 eV$. The resulting energy gap is $E_{gap} = 2.2 eV$. The following investigations will concentrate on pentacene thin films deposited by physical vapor deposition. The utilized evaporation chambers are described by section 3.2.

The applied material was obtained from Sigma Aldrich as fluorescence grade. However, since the purity of the organic semiconductor is critical for its charge transport properties, the original material was gradient sublimed twice before processing. For this process, a tube furnace at a maximum temperature of 260°C, a negative temperature gradient of 3.9 $\frac{°C}{cm}$ as well as a constant stream of Ar / H_2 gas at 0.7mbar was utilized.

---

[8] Molecular dimensions of pentacene: a=7.9Å, b=6.06Å and c=16.01Å.

3.2 Sample preparation

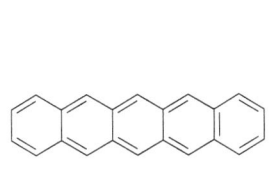

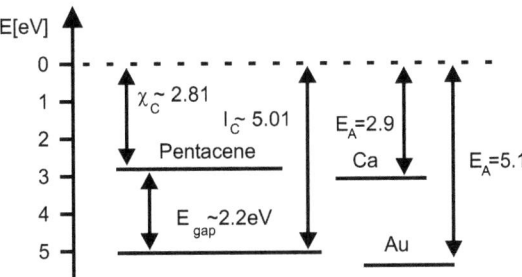

(a) Molecular structure of pentacene (C22H14)

(b) Comparative energy diagram between the pentacene transport states I_C (HOMO) and χ_C (LUMO), as well as the work function (E_A) of the utilized source-drain metalizations [31, 67].

Figure 3.2

Metals

The metals selected for the following investigations are Au and Ca, with a respective purity of 99.99% (Chempur) and 99.5% (Alpha Aesar). These materials were chosen due to their work function of 2.9eV for Ca and 5.1eV for Au [67], as comparatively illustrated to the HOMO / LUMO levels of pentacene in Figure 3.2(b). The work function matching of these metals to the respective transport levels of the organic semiconductor ideally results into almost ohmic contacts[9] for the injection of electrons (Ca) and holes (Au).

3.2 Sample preparation

Section 3.2 provides the necessary details for the sample preparation. In a first subsection, the equipment used for the thin film deposition is listed, followed by a discussion of the process sequence for the sample realization. In a final subsection, several sample conditioning steps, which are used to influence the charge carrier transport properties of selected devices, are deliniated.

3.2.1 Thin film deposition

Thin films by physical vapor deposition

The physical vapor deposition of metal or organic thin films is conducted by the use of two Balzers evaporation chambers. The materials are evaporated, using current heated crucibles. The utilized crucible material for the respective evaporated matter is summarized in Table 3.2. The deposition chambers are connected by a transfer system, allowing for the subsequent evaporation of semiconductors and metals without breaking the vacuum. Furthermore, a docking possibility for an available transfer shuttle allows for the transport of the samples to and from the PVD-system to the respective

---

[9] Here ohmic contacts are considered as: Contacts that allow charge carrier injection without contact resistance during device operation.

Crucible	Sublimation Temperature [°C]	Material	Sublimation Temperature [°C]
Ta	3200	Au	1400
Mo	-	Ca	-
SiO$_2$	2000	C$_{22}$H$_{14}$	252 [70]

Table 3.2: Summary of the evaporated materials and their corresponding crucibles. The sublimation temperature of the materials is specified for a pressure of 10^{-2} mbar.

preparation or measurement glove boxes without exposure to ambient air. The influence of the available laboratory environment, with respect to the device performance, can therefore be excluded. For more detailed information on the Balzers evaporation system, the reader is referred to the dissertations of Heil [68] and Hepp [69].

In order to structure the thin films during their deposition, shadow masks with a thickness of 100μm are positioned in the sample holders prior to their introduction into the vacuum of the evaporation chambers. The stainless steel masks were obtained from CADiLAC-Laser. Thin polymer spacers were used in between the sample and the metal shadow mask.

Thin films by spin coating

In order to deposit polymers from solution, a Specialty Coating Systems spin coater (Model: P-6708D) located in a MBraun Unilab glove box, with an inert H$_2$O atmosphere (N$_2$ ≤ 1ppm, O$_2$ ≤ 1ppm) was utilized. As described above for the Balzers evaporation system, the glovebox system has a transport shuttle docking possibility, in order to be able to transport the prepared samples without exposure to air. The exact dilutions and spin coating parameters for the processed polymers are summarized in Appendix D (Tables D.1 and D.2).

3.2.2 Sample structure

In the following, an overview of the process sequence for samples investigated by chapters 4-6 is given. The exact process details, however, are described in Appendix D. All of the described physical vapor deposition steps were conducted at a chamber base pressure <10^{-6}mbar.

The standard substrate used for all of the processed samples is a p^{++} doped 17x17 mm^2 silicon substrate with a 200nm dry oxide, as grown by the supplier (ChemPur). In a *first process step*, the substrates are cleaned by subsequently sonicating the samples in a 5% vol. dilution of deconex (Borer Chemie) and deionozed water, and then in pure deionized water for time frames of 15 minutes. The substrates are dried in a stream of pure N$_2$.

In dependence of the sample application, the *second process step* consists either of the deposition of a thin layer of Ca or of the deposition of different polymeric insulators. The Ca layer is evaporated onto the SiO$_2$ insulator by PVD for a thickness range between 0.6Å and 250Å at a deposition rate of $0.4\frac{\text{Å}}{s}$. The Ca layer is structured during its deposition by the use of a shadow mask. The deposition of the polymers from solution is conducted by the use of a spin coater. The molecular structure of the

3.2 Sample preparation

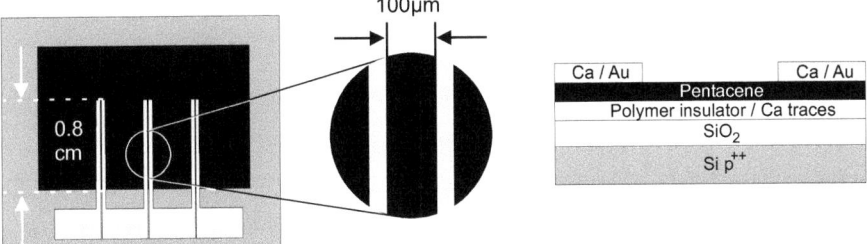

(a) Top view of the transistor structure with a $\frac{w}{l}$ ratio of 80.

(b) Schematic cross section of the transistor structure.

Figure 3.3: Schematic illustration of the used top contact transistor structure geometry.

utilized polymers is illustrated in Figure 3.1. The respective dilutions, the exact spin coating parameters, as well as the resulting layer thicknesses ranging from d =119nm to d =212nm are described by Appendix D.

The *third process step* is optional, and is only applied in case of the deposition of a polymer insulator onto the SiO$_2$ dielectric. For this step, the processing of the samples in inert N$_2$ atmosphere or vacuum is interrupted, in order to expose the polymer layer to UV radiation in ambient atmosphere. The UV exposure is conducted at wavelengths of 254nm and 185nm, with a respective optical power of 15mW and 1.5mW, using the ozone photoreactor PR-100 (Ultra Violet Products).

The deposition of a 50nm pentacene layer is realized in a *fourth process step* by the use of PVD at a deposition rate of $2\frac{\text{Å}}{s}$. For the case of deposited Ca traces (process step two), the pentacene layer is deposited directly after the Ca deposition onto the same area without breaking the vacuum.

In a final and *fifth process step*, the source-drain metalization of either Au or Ca (d = 100nm) is deposited by PVD at a deposition rate of $2\frac{\text{Å}}{s}$. The source-drain geometry, and therefore the channel dimensions of w = 0.8cm (channel width) and l = 100μm (channel length), are defined during the deposition by the use of a shadow mask.

As already indicated by the discussed process sequence, a top source-drain / bottom gate contact transistor structure is investigated. The cross section of the transistor structure as well as the transistor channel dimensions are illustrated in Figures 3.3(a) and 3.3(b). In order to modify or to characterize the individual layers, the process sequence is interrupted after the respective process steps, as described in the following.

3.2.3 Electrical characterization and sample conditioning

In the following section, the periphery for the electrical device characterization as well as several sample conditioning methods by which the current voltage characteristic of the investigated OFETs is influenced, are introduced. The term *conditioning* defines the exposure of OFET devices to electrical stress or a thermal charging (forming) step.

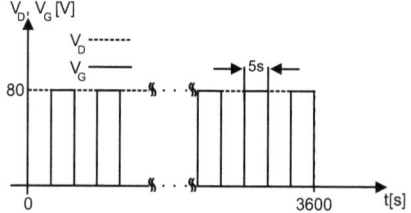
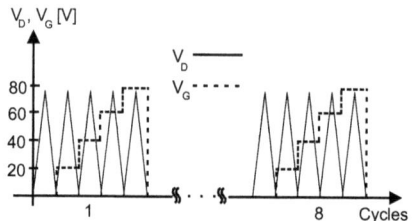

(a) Electrical conditioning applied to OFETs incorporating a Ca passivation layer. Here, V_D is held constant at 80V, while V_G is pulsed to 80V in 5s intervals for the duration of 1h.

(b) Electrical conditioning applied to OFETs incorporating a UV modified PMMA dielectric. For this approach the output characteristic is driven for 8 cycles in the electron accumulation. For each cycle V_G is varied between 0V and 80V in $\Delta V_G = 20$V steps, and V_D is varied between 0V and 80V by $\Delta V_D = 1$V steps for each V_G step.

Figure 3.4: Driving schematic used for the different electrical cyclic conditioning steps.

Electrical characterization

The electrical characterization of OFETs is conducted by the use of an HP 4155A parameter analyzer, as discussed in the following chapters. For specific applications, such as the characterization of complementary metal oxide semiconductor (CMOS) inverter structures, or one of the electrical cyclic conditioning steps, as considered in the following subsection, a Keithley 6517A electrometer is utilized in addition to the parameter analyzer.

Electrical cyclic conditioning

The type of electrical conditioning is dependent on the OFET cross section. In the following, two cyclic conditioning steps are introduced. The first approach is only applied to OFET samples incorporating a Ca passivation layer on the SiO$_2$ dielectric, while the second approach is exclusively applied to OFET devices, incorporating an additional PMMA dielectric exposed to UV radiation in ambient atmosphere.

1. It is demonstrated in section 5.1.4, that the electrical OFET performance of devices incorporating a *Ca passivated SiO$_2$ dielectric* can be significantly improved, if the transistors are exposed to electrical stress in inert N$_2$ atmosphere. The cyclic conditioning is applied in the electron accumulation mode for the duration of one hour. During this time frame, the drain voltage is held constant at $V_D = 80$V and the gate voltage is pulsed in 5s intervals between $V_G = 0$V and $V_G = 80$V [54]. The pulse diagram for this type of cyclic electrical conditioning is illustrated in Figure 3.4(a).

2. The cyclic electrical conditioning applied to OFETs incorporating a *UV modified polymer dielectric* is conducted by driving the transistor output characteristic for eight cycles in the electron accumulation mode. For each cycle, the gate voltage is increased in $\Delta V_G = 20$V steps from $V_G = 0$V to $V_G = 80$V. For each step, the drain voltage is applied by scanning the drain

3.2 Sample preparation

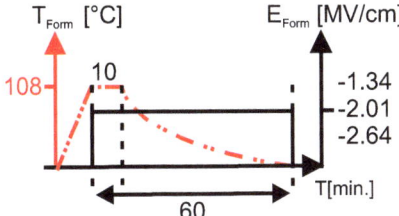
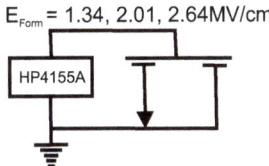

(a) Time sequence of forming temperature and field.

(b) The forming was conducted in individual forming steps.

Figure 3.5: PMMA dielectric forming setup and equivalent circuit.

voltage from $V_D = 0V$ to $V_D = 80V$ and back to $V_D = 0V$ in $\Delta V_D = 1V$ intervals. The pulse diagram of this conditioning step is depicted in Figure 3.4(b). The influence of this type of electrical stress with respect to the transistor performance is delineated in section 5.2.2.

Thermal charging under the influence of an electric field

The influence of a thermal forming step (see section 2.3), on the electrical performance of an OFET, incorporating a SiO_2 / PMMA dual layer dielectric, will be discussed in detail in chapter 6. In the following, the implemented forming scheme, with regard to timing and applied field strengths as well as the utilized experimental setup is delineated.

In order to charge the PMMA dielectric in a completed OFET device structure, and to investigate the influence of the charging on the OFET threshold, different forming field strengths E_{Form} (F_F= -1.34$\frac{MV}{cm}$, -2.01$\frac{MV}{cm}$ and -2.64$\frac{MV}{cm}$) are applied between the gate electrode and the source-drain contacts. The forming step was conducted at a temperature of T_F = 108°C. The process sequence of the subsequent forming steps as well as the equivalent circuit for the application of the forming fields is illustrated in Figures 3.5(a) and 3.5(b). The individual forming fields are applied to the transistor for a duration of 60 minutes. For the first 10 minutes of the forming step, the temperature is held constant at T_F. Subsequently, the temperature is allowed to cool down to room temperature during a time frame of 50 minutes, while the forming field is held constant at the respective field strength.

The experimental setup used for the thermal charging is illustrated in Figure 3.6. A RCT basic Hotplate (IKA Werke) is used as heat source. In order to electrically isolate the grounded hotplate from the bottom gate of the transistor, a sheet of teflon is placed on top of the heating surface. The copper block, on top of the teflon sheet, is used to contact the gate electrode of the transistor, and to gain temperature stability during the forming step. The temperature of the substrate is obtained by the use of a flat head type C thermocouple, connected to a multimeter (HGL 3300). After placing the OFET on

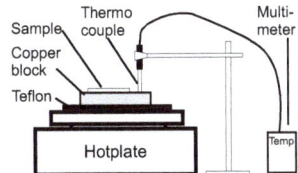

Figure 3.6: Thermal forming step experimental setup, excluding the HP4155A parameter analyzer.

top of the copper block, the source-drain electrodes are connected to a HP4155A parameter analyzer by means of MH4-B (Cascade Microtech) micro probes. The gate electrode is connected to the parameter analyzer by the use of the copper block and an additonal micro probe. The parameter analyzer is utilized to apply the electric field strength and to record the displacement current during the experiment.

3.3 Thin film characterization

3.3.1 Atomic force microscopy

The atomic force microscopy (AFM) measurements, as discussed in chapter 5.2.1, were conducted by using an Atomicprobe CP microscope (Park Scientific Instruments) and NSC11 / 50 cantilevers (MikroMasch) with a respective tip radius and angle of 10nm and 20°. All measurements were conducted in air by the use of the non-contact mode. In order to investigate the AFM images, the software Gwyddion 2.9 (Open Source Software) was utilized. Prior to the analysis, the images were flattened, to compensate for scanning artifacts during the measurement.

3.3.2 Water contact angle measurement

As discussed in chapters 4.2 and 5.2.1, water contact angle measurements were conducted by the use of a Krüss G10 contact angle measurement system operated in air. The system includes a G1041 video setup, which can be accessed by the Krüss Drop Shape Analysis software.

In general, water contact angle measurements are very sensitive with regard to the experimental conditions. The measurement is dependent on such parameters as water drop size, measurement time, temperature, adsorbates and surface roughness [71, 72]. However, for the work at hand, the surface roughness of the samples can be neglected, due to values of $< 1 \mu m$ [73]. To keep the experimental conditions similar for all of the conducted measurements, the following measurement protocol was followed:

A $0.6 \mu l$ drop volume of deionized water is deposited onto the sample surface. The respective measurements are taken 75s after the deposition of the water droplet. The water contact angle values, as specified in the following chapters, represent the mean value of two or three measurements per sample. For these measurements, the deviation in contact angle never exceeded ±3°.

3.3.3 Layer thickness determination

To obtain the layer thickness of deposited thin films, a Dektak IIa (Veeco) profilometer as well as a SE850 (Sentech) spectral ellipsometer are utilized in dependence of the investigated material. Due to the strong interatomic bonds of metals and their resulting insensitivity to the bearing forces of a profilometer, the Dektak IIa is used for the thickness characterization of metalizations. For organic thin films, however, the interatomic binding forces of molecular solids are defined by weak van der Waals interaction[10], making the measurement of the layer thickness by the use of a profilometer

---

[10] Van der Waals interaction exhibits binding energies in the order of $\approx 0.1 eV$ [74].

3.3 Thin film characterization

prone to errors. Despite the weak binding forces of small molecule thin films, the Dektak IIa could be applied for thickness measurements of such layers, as verified using the mechanical force free spectral ellipsometer. To obtain the layer thickness of polymer thin films, however, the Sentech spectral ellipsometer had to be applied, in order to avoid significant measurement errors.

3.3.4 Photoelectron spectroscopy

In the following section, the basic principles for the understanding of the photoelectron spectroscopy (PES) measurements, as discussed in chapter 5, are summarized. For further information, the reader is referred to more in-depth literature [75–77].

The *principle of photoelectron spectroscopy* is based on the external photoelectric effect, by which photoelectrons are excited to escape the investigated sample using high energy radiation. For the case of X-Ray photoelectron spectroscopy (XPS), the radiation energy exceeds 1000eV [75]. This type of radiation allows for the photoionisation of atomic core levels and thereby permits the investigation of the chemical composition of the sample by an energy discriminative analysis of the emitted photoelectrons. This is possible, since each element shows a unique set of orbitals with binding energies in dependence of its chemical environment. The investigated emission spectra are mainly composed of primary, secondary and Auger electrons emitted from the sample. However, since this dissertation focuses on the analysis of the chemical composition of the various samples, a discussion of Auger electrons is neglected in the following. For the case at hand, primary electrons, which have not experienced an energy loss while leaving the sample, are the most useful [76] for a chemical analysis. These electrons are responsible for the distinctive emission lines of the spectrum. The low energy background of the spectrum is formed by photoelectrons which have experienced energy losses by inelastic scattering during the emission process, as well as secondary electrons (excited valence electrons), which are excited by inelastic scattering. Due to the high interaction probability of electrons with matter, the mean free path length (λ_m) for photoelectrons, and therefore the emission of primary photoelectrons, is limited to a few tens of angstroms in dependence of their kinetic energy[11]. Photoelectron spectroscopy is therefore a surface sensitive method.

A photoelectron is able to escape the sample surface for radiation energies $h\nu$ larger than its binding energy $E_B^V(k)$. The binding energy with reference to the vacuum level of an electron located in core level k is given by the difference in energy between the atom in its final state (E^f) and its initial state (E^i) prior to the excitation. However, for photoelectron spectroscopy on solids, the binding energies of the electrons are generally measured with respect to the Fermi level (E_F) [75] by determining their kinetic energy E_{kin}, using an energy discriminator. Therefore, the binding energy E_B^F can be calculated in accordance with the following equation.

$$E_B^F = h\nu - E_{kin} - \phi_s \qquad (3.1)$$

In this equation ϕ_s represents the sample work function. Due to an electrical connection between the sample and the spectrometer during the measurement, their respective Fermi levels are aligned. For a

---

[11] In general, the photoemission is limited to $3*\lambda_m$, where 97% of the photoelectrons are attenuated.

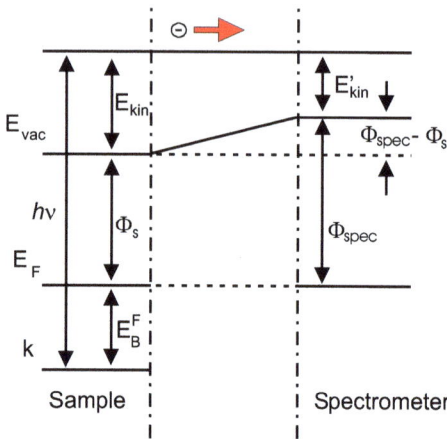

Figure 3.7: Schematic energy level diagram during the photoelectron excitation and detection [78]. The binding energy is measured with respect to the work function of the spectrometer. A photoelectron has to surpass the energy difference between the sample and spectrometer work functions ($\Phi_{spec} - \Phi_s$).

possible difference in work function between the sample and the spectrometer, as indicated in Figure 3.7, the kinetic energy of the photoelectron is altered by this difference, as defined by equation 3.2.

$$E_B^F = h\nu - E_{kin} - \phi_s - (\phi_{spec} - \phi_s) \tag{3.2}$$

Here, the work function of the spectrometer is represented by ϕ_{spec}. Therefore, the obtained photoelectron binding energies E_B^F are independent of the sample work function and are always measured with regard to the work function of the spectrometer.

The photoemission of an electron produces an atomic final state, that is lacking one electron compared to its initial state. Therefore, only photo hole state energies can be obtained by the use of PES. Such final states differ from their neutral ground states, for example, by the relaxation of the atomic shell, which may lead to a split of distinctive core level emission lines into dubletts due to spin orbit coupling. This kind of final state effect is the result of the interaction between spin and angular momentum of unpaired electrons. The resulting total momentum j is obtained by a superposition of the angular momentum l and the spin momentum s ($j = l \pm s$) [79]. The possible spin orbit coupling for s- and p-orbitals[12] is illustrated in Table 3.3. Due to an angular momentum of $l = 0$, the total momentum of an unpaired electron in the s-orbital always exhibits a value of $j = \frac{1}{2}$. Therefore, with respect to PE spectra, emission lines, resulting from s-orbitals, are generally not subject to spin orbit splitting. However, due to an angular momentum of $l = 1$ for electrons located in a p-orbital, the total momentum is either given as $j = \frac{1}{2}$ or $j = \frac{3}{2}$. This results into a spin orbit splitting of the PE emission lines (dubletts), with their components being separated by a delta in energy in dependence

---

[12] In general the spin orbit coupling is described by the "$L - S$" coupling approximation for light atoms, and by the "$j - j$" coupling approximation for heavy atoms [79]

3.3 Thin film characterization

Orbital	l	s	$j = l \pm s$	$\frac{2j^-+1}{2j^++1}$
s	0	$\frac{1}{2}$	$\frac{1}{2}$	-
p	1	$\frac{1}{2}$	$\frac{1}{2}, \frac{3}{2}$	1:2

Table 3.3: Angular (l), spin (s) and total momentum (j) of the atomic s- and p-orbitals. In addition, the intensity ratio ($\frac{2j^-+1}{2j^++1}$) for possible dublett states of photoelectrons is indicated.

of the atomic number. As shown in Table 3.3, the intensity ratio of 1:2 for the respective components represents the ratio of the number of possible degeneracies for these components.

The analytical power of PES is given by the fact that the measured photoelectron binding energy is dependent on the chemical environment or oxidation state of the respective atom. This initial state effect is called a chemical shift and is the result of a change in the partial charge of the atom, which leads to an in- or decrease in the photoelectron binding energy. Such changes may be due to the gain or loss of one or more electrons by oxidation in ionic binding or a partial charge due to the shift in the charge centroid of a covalent bond between atoms exhibiting a different electronegativity.

Furthermore, the binding energy is influenced by charging effects of the investigated samples. If the lifetime of the photo hole, caused by the emission of a photoelectron, is in the order of the measurement time, possibly due to a low conductivity of the investigated sample, the sample exhibits a positive charge, which influences the binding energy by a shift to higher values. Such effects are usually observed for the investigation of insulators or thin films, deposited onto insulating substrates and have to be compensated by an electron flood gun.

The spectral line width of a photo emission line ΔE is influenced by the natural line width ΔE_n of the transition between E^i and E^f, due to the lifetime t_n ($\Delta E_n * t_n \leq \hbar$) of the excited photo hole, the spectral line width of the excitation radiation ΔE_p as well as the resolution ΔE_d of the spectrometer. This is described by the following equation:

$$\Delta E(FWHM) = \sqrt{\Delta E_n^2 + \Delta E_p^2 + \Delta E_d^2} \qquad (3.3)$$

The **equipment** used for the XPS measurements is the DArmstadt Integrated SYstem for MATerial science, in short **DAISY-MAT**. This system allows for the sample preparation, such as the deposition of thin films by PVD as well as the subsequent analysis of the samples without breaking the UHV ($\approx 10^{-10} mbar$). The XPS unit PHI5700 (Physical Electronics) contains monochromatized Al Kα and Mg Kα radiation sources as well as the PHI electron analyzer (spectrometer). The specifications of the analyzer unit, with respect to the utilized radiation energy, the resulting resolution of the radiation sources in combination with the electron analyzer, as well as its used angle of incidence perpendicular with regard to the sample are summarized in Table 3.4. To obtain the electron analyzer work function (ϕ_s), which is required for the calculation of the photoelectron binding energy, as indicated by equation 3.2, the XP-spectrum of a clean metal sample is obtained, prior to each measurement. By considering the Fermi edge of the metal sample, the electron analyzer is calibrated with reference to this value.

Radiation Source	E($h\nu$) [eV]	Resolution [meV]	Angle of incidence Θ [°]
Al Kα Mono	1486.6	< 500	45
Mg Kα Standard	1253.6	< 900	45

Table 3.4: Specifications of the analyzer unit PHI5700 (Physical Electronics).

In the following, the basic principles for the ***analysis of the obtained PE spectra*** are discussed:

- The assignment of photoemission lines to the chemical species of an investigated sample is obtained by comparing the experimental data to databases such as the Handbook of X-ray Photoelectron Spectroscopy [76] or the NIST X-ray Photoelectron Spectroscopy online database [80]. Their intensity values are given by the integrated countrate of an emission line, and are represented in the spectrum by the peak area enclosed by the emission. The intensities are obtained by first subtracting the background of the spectrum, using a Shirley function [81], and then by fitting the respective components, using a Voigt function. A Voigt function is the result of the convolution of a Gaussian and a Lorenzian function. The Gauss function represents the emission line broadening, due to the excitation radiation as well as the spectrometer, whereas the natural line width of the photo hole is considered by the Lorenzian function. In addition to the determination of the emission line intensity, this approach allows for the deconvolution of the envelope curve of several emission lines, located close to each other, e.g. due to a chemical shift, with respect to their binding energy.

- To quantify the chemical composition of the investigated sample, a method, using peak area atomic sensitivity factors (S), can be applied to obtain the relative concentrations of the various constituents. This approach is valid under the assumption of a stochastically homogeneous thin film.

$$C_x = \frac{n_x}{\sum n_i} = \frac{I_x/S_x}{\sum I_i/S_i} \quad (3.4)$$

The relative concentration of a sample component C_x can be obtained by weighing its intensity I_x with the respective atomic sensitivity factor S_x (ASF) in relation to the sum of all weighted sample components (equation 3.4). The atomic sensitivity factor is, amongst other parameters, dependent on the photoelectric cross-section of the atomic orbital of interest, as well as an angular efficiency factor with respect to the angle α between the photon path and the detected electron path [13]. The ASFs used in chapter 5 for carbon and oxygen are C1s = 0.296 and O1s = 0.711.

- Under the assumption of a homogeneous thin film, its layer thickness can be obtained using PES, by considering the following equation:

$$\frac{I_{Sub}^d}{I_{Sub}^0} = exp(-\frac{d}{\lambda_m(E)sin\Theta}) \quad (3.5)$$

---

[13] $\alpha = 90°$ for the work at hand

Here, Θ symbolizes the angle between the horizontal plane of the substrate and the electron analyzer. Equation 3.5 represents a correlation between the intensity I_{Sub} of the substrate photoemission line, the mean free path length $\lambda_m(E)$ of the photoelectron as well as the thickness d of the deposited thin film. By measuring the intensity of the substrate emission lines, prior to (I^0_{Sub}) and after the deposition of the investigated thin film (I^d_{Sub}), its thickness can be derived by conversion of equation 3.5.

Chapter 4

Ambipolar/Unipolar OFET charge carrier transport

As indicated in chapters 2.2.2 and 3.1, the OFET charge carrier transport properties can be influenced by the dielectric / semiconductor interface or by the choice in source-drain metalization. In this chapter, it will be argued, how to define the OFET polarity as unipolar n- / p-type or ambipolar by the use of an adequate source-drain metalization. Furthermore, the influence of different polymer gate dielectrics on the unipolar OFET charge carrier transport will be investigated. For this type of discussion, a semiconductor, preferably with balanced ambipolar charge carrier transport properties, is essential. In the following, the organic semiconductor pentacene is used as a model semiconductor. Even though pentacene electron conduction was demonstrated by Minakata et al. as early as 1993 by the use of alkaline metal doping [82], pentacene was until recently assumed to exhibit exclusive intrinsic p-type charge carrier transport properties [20]. However, this view was disproved by Meijer et al. (2003, [55]), who were able to demonstrate the first electron field effect mobility for pentacene OFETs. Illustrated in Figure 4.1 are the maximum pentacene charge carrier mobilities, published for the time span between 1992 and 2007, pertinent to hole as well as electron transport. The respective references are listed in Table C.2. While at first the published data for ambipolar charge carrier transport in pentacene was still unbalanced ($\mu_e = 10^{-6} \frac{cm^2}{Vs}$, $\mu_h = 10^{-2} \frac{cm^2}{Vs}$), Ahles et al. [28, 83] succeeded in demonstrating balanced n- and p-type field effect charge carrier mobilities for pentacene OFETs in the order of $0.1 \frac{cm^2}{Vs}$.

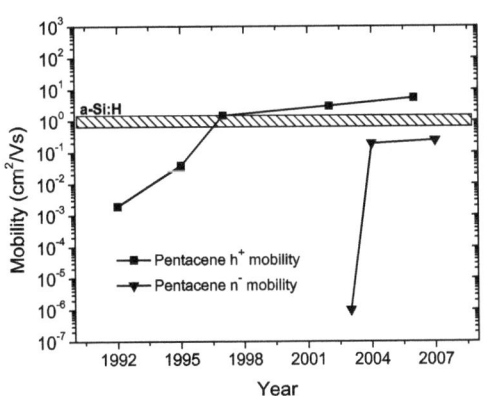

Figure 4.1: Development in OFET pentacene n / h mobility between 1992 and 2007. The shaded bar represent the a-Si:H mobility range [11].

4.1 OFET polarity in dependence of the source/drain metalization

The experiments were performed using pentacene OFETs, realized on a p^{++} – Si substrate, comprising a 200nm dry oxide. However, as demonstrated by section 2.2.2, electronic states at the dielectric / semiconductor interface, such as hydroxyl groups [14], play a vital role with regard to the OFET charge carrier transport. Therefore, in order to enable both electron and hole charge carrier transport in pentacene, an additional PMMA dielectric, without hydroxyl groups in its chemical structure, is deposited on top of the SiO_2 interface (d_{PMMA} = 119nm) [84]. The pentacene layer as well as the source-drain metalization are deposited by PVD with a respective thickness of 50nm and 100nm. The applied $\frac{w}{l}$ ratio of the source-drain metalization is 80. For more information on the standard OFET device structure or process conditions, the reader is referred to the experimental section (chapter 3.2).

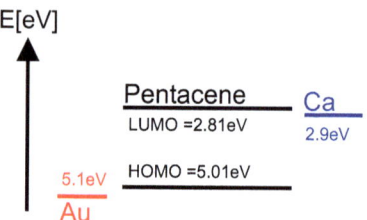

Figure 4.2: Work function of Au and Ca relative to pentacene HOMO / LUMO levels.

By matching the work function of the source and drain metalization to the HOMO or LUMO level of the organic semiconductor, as illustrated in Figure 4.2, the polarity of an OFET can be defined as unipolar n- or p-type. This is the consequence of a low injection barrier for one charge carrier type, while the injection barrier for the complementary charge carrier type is high. For the case of an Au / pentacene interface, this translates into an almost ohmic contact for the hole injection into the pentacene HOMO level as well as a large injection barrier of ≈ 2.3eV for the injection of electrons into the LUMO level of the organic semiconductor. The resulting p-type transfer characteristic for such an OFET is demonstrated by the open triangle scatter plot, illustrated in Figure 4.3. Using this characteristic in the hole accumulation, a hole mobility value of $\mu_h = 1.2 * 10^{-1} \frac{cm^2}{Vs}$, a threshold voltage of $V_{th,p} = -19.2V$ as well as an $\frac{On}{Off}$ ratio of ≈ 10^4 is derived. The complementary experiment, using a Ca source-drain metalization, results in a unipolar n-type device, due to the matching of

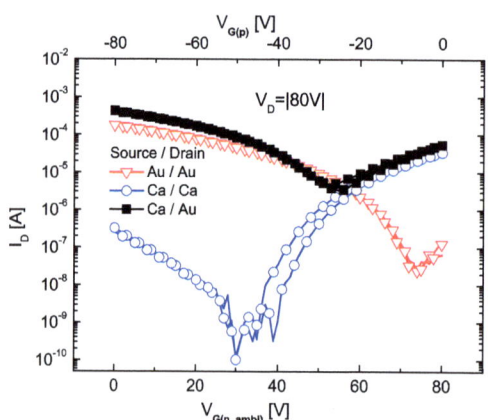

Figure 4.3: Transfer characteristics of pentacene OFETs, differing only in the source-drain metalization. By matching the metal work function to the respective pentacene HOMO / LUMO levels, the transistor polarity is defined as either unipolar n- / p-type or ambipolar.

---

[14] Hydroxyl groups are available on SiO_2 interfaces with an area density of $3 - 7 * 10^{13} \frac{1}{cm^2}$ [29].

the metal work function to the pentacene LUMO level and the hence associated respective low/ high ($\approx$2.1eV) injection barriers for electrons / holes. The transfer characteristic of this device in the electron accumulation is illustrated by the open dot scatter plot in Figure 4.3. Using this characteristic, an electron mobility of $\mu_e = 7.8 * 10^{-2} \frac{cm^2}{Vs}$, a threshold voltage of $V_{th,e} = 46V$ as well as an $\frac{On}{Off}$ ratio of $\approx 10^4$ is extracted. In addition to the expected current profile of the n-type and p-type characteristics, the OFETs exhibit an increase in drain current for low gate voltages. This experimental artifact is the result of a leakage current between the drain and the gate electrode, as determined by considering the gate leakage current, recorded in parallel to the measurement.

An ambipolar OFET, where both electrons and holes are injected and transported in the transistor channel, can be realized by using a Ca source as well as a Au drain metalization. The transfer characteristic of such a device in the electron accumulation is illustrated by the filled square scatter plot in Figure 4.3. The ambipolar transfer characteristic can be described by the extended Shockley equations, as introduced in chapter 2.2.1. For low gate voltages ($|V_{G,ambi} - V_D| \approx |V_D| \geq |V_{th,h}|$), positive charge carriers are injected into the transistor channel by the Au drain electrode, due to the negative potential difference between the gate and the drain contact as well as the low injection barrier at the Au / pentacene interface. This results in the observed substantial drain current. However, with an increase in gate voltage, the drain current is reduced, due to a reduction in potential difference between the drain and the gate electrode. For $V_{G,ambi}$, exceeding the threshold voltage for electrons ($V_{th,n}$), the electron injection from the source electrode results once more in an increase in drain current,that converges toward the transfer characteristic of the unipolar Ca-Ca contact n-type device. The electron and hole mobility values, as well as the respective threshold voltages derived from this transfer characteristic, are in agreement with the values obtained, using the respective unipolar current voltage characteristics. An exception is the $\frac{On}{Off}$ ratio, which is ill defined for ambipolar devices, due to the conduction of both charge carrier types in the transistor channel. In contrast to the unipolar devices, no non-conductive state for the ambipolar transistor is obtained. Thus the $\frac{On}{Off}$ ratio is meaningless. The unipolar n-, p-type and ambipolar characteristics exhibit a low current hysteresis, which is suggested to be the result of a low density of energetically shallow charge carrier traps at the dielectric / semiconductor interface. However, the presence of energetically deep traps cannot be excluded, especially when considering the high threshold voltages for electrons and holes. Illustrated in Figure 4.4 is the output charac-

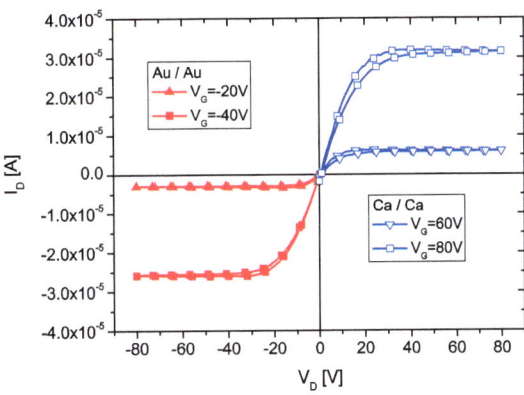

Figure 4.4: Unipolar output characteristics in the hole and electron accumulation. The filled and open scatter plots represent the respective unipolar p-type and n-type characteristic of a Au / Au or Ca / Ca electrode pentacene OFET.

teristic of the unipolar n- and p-type OFETs in the respective electron and hole accumulation modes. These characteristics support the suggested low injection barrier for holes from Au electrodes and for electrons from Ca electrodes into the semiconductor. This is deducted from the absence of an S-shaped current behavior in the linear range of the respective characteristics.

In this section it was shown, that unipolar n- / p-type or ambipolar pentacene OFETs can be realized by the choice in source-drain metalization as well as an appropriate gate insulator. Due to the matching of the metal work function to either the HOMO or LUMO level of the semiconductor, a respective injection of either holes or electrons can be achieved. This is the result of a low injection barrier for one charge carrier type, while the injection of the complementary charge carrier type is blocked by a high injection barrier. Therefore, by the use of different metals for the source and drain metalization, an ambipolar OFET can be realized. These conclusions hold true on the condition of an ambipolar organic semiconductor as well as a gate dielectric, which allows for balanced charge carrier transport in the semiconductor.

4.2 Influence of different dielectrics on OFET charge carrier transport

In the following section, the influence of several polymer gate dielectrics, with varying amounts of oxygen containing polar groups in their repeating chain, is investigated with regard to unipolar p- and n-type charge carrier transport properties of pentacene OFETs [44]. The chemical structure of the utilized polymers, which are PS, PC, PMMA, P4VP and PI are depicted in Figure 3.1. The dielectric constant of the investigated polymers as well as the respective water contact angles (97°, 92°, 81°, 76° and 67°) are summarized in Table 3.1.

The OFET structure, used for the following experiments, is the same as described in section 4.1, however, it differs as to the utilized polymer gate dielectric. The respective polymer dilutions, as well as the resulting layer thicknesses obtained by spin coating, are specified in Appendix D (Table D.1). The OFET polarity is defined as either unipolar p- or n-type by the use of an Au or Ca source-drain metalization.

Depicted in Figure 4.5(a) and Figure 4.5(b) are the OFET threshold charge carrier density n_{th} and mobility μ for electrons ($n_{th,e}$, $\mu_{th,e}$) and holes ($n_{th,p}$, $\mu_{th,p}$) as a function of the polymer dielectric contact angle. In the experimental section, the decrease in water contact angle for the different polymers was suggested to be the result of an increase in the amount of oxygen containing polar groups in their corresponding monomeric units and therefore the dielectric interface. The error bar statistic was conducted, utilizing up to 4 samples out of different production batches. A maximum of 3 OFETs per sample was considered. The parameters were determined, by using the respective transfer characteristics in the electron and hole accumulation mode. Here, the areal charge carrier density n_{th} is used instead of the threshold voltage V_{th}, in order to account for differences in the total device capacitance. The threshold charge carrier density is calculated from the following equation:

$$n_{th} = C_{tot} * \frac{|V_{th}|}{|e|} \quad (4.1)$$

4.2 Influence of different dielectrics on OFET charge carrier transport

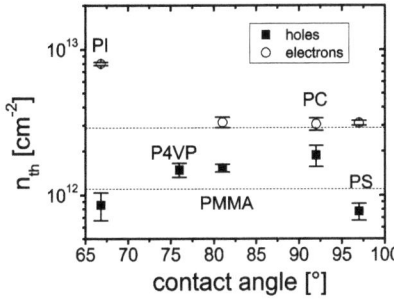
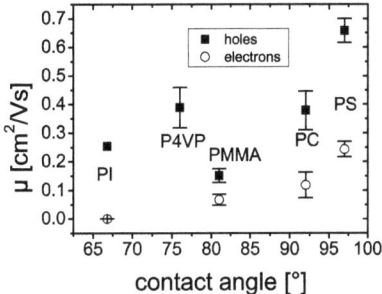

(a) n_{th} for electrons and holes in dependence of the water contact angle. The dotted lines represent orientation lines

(b) μ for electrons and holes in dependence of the water contact angle

Figure 4.5: n_{th} and μ in dependence of the water contact angle. For the error bar statistic up to 4 samples out of different batches were utilized, with a maximum of 3 OFETs per sample.

The elementary charge is defined by e and C_{tot} represents the total device capacitance of the SiO$_2$ / polymer double layer dielectric.

By considering Figure 4.5(a), an increase in $n_{th,e}$ with a decrease in contact angle is found, while no clear correlation between a change in contact angle and $n_{th,h}$ is obtained. This is suggested, since a decrease in water contact angle, due to an increase in oxygen containing polar groups, available at the dielectric interface, is expected to lead to a broadening of both the HOMO and LUMO DOS of the semiconductor, due to dipolar disorder [52]. This should result in a degradation of the transistor transport performance for both electrons and holes. Since, however, only a degradation in $n_{th,e}$ is obtained, it is suggested, that oxygen containing polar groups, available in the form of Keto and hydroxyl groups, represent electron traps, energetically positioned in the band gap of the semiconductor. This is supported in Figure 4.5(b). Here, as for the case of n_{th}, no clear correlation between the hole mobility μ_h and available Keto and hydroxyl groups can be derived, while the electron mobility μ_e is improved with a decrease in the amount of oxygen containing polar groups at the dielectric interface. The negative influence of hydroxyl groups on the charge carrier transport seems to be much stronger than the influence of Keto groups. This is demonstrated for the case of P4VP, where the presence of one hydroxyl group per monomer unit completely inhibits the electron charge carrier transport. This is supported by the fact, that PI, which exhibits a lower contact angle as well as a higher concentration of Keto groups in its monomer unit when compared to available hydoxyl groups in the monomer unit of P4VP, still allows for OFET electron transport with a low electron mobility of $\mu_e = 3 * 10^{-4} \frac{cm^2}{Vs}$. The trapping nature of Keto groups, as well as the pronounced electron trapping nature of hydroxyl groups, is in agreement with the findings of others [29, 43]. However, only limited information about possible electron trapping mechanisms or energetic trapping depths for Keto and hydroxyl groups is available. Kadashchuk et. al [43] have estimated an energetic electron trap depth of 0.64eV for Keto groups, available in a photooxidized polyfluorene derivative, using thermally stimulated current (TSC) measurements. However, no trapping mechanism was proposed. As to hydroxyl groups, an electron trapping mechanism by dissociation of a hydrogen atom has been presented by Chua et al. [29]. Hence, a recharging of a discharged trap should not be possible. However, experimental evidence, presented in section 5.2.2, indicates, that the discharge process of electrons, trapped by hydroxyl groups, is reversible, questioning the proposed decomposition of the OH-groups.

The influence of morphology on the charge carrier transport properties has not been considered by the discussion so far. While the semiconductor morphology may influence the OFET transport properties, as observed for the case of holes by Karl et al. [85], the conducted experiments unveil no clear dependence of the hole mobility on the different water contact angle values of the investigated dielectrics and hence on potentially changing pentacene morphologies. Since the influence of the semiconductor morphology is expected to be similar for the transport properties of electrons and holes, the observed dependence of the electron mobility is essentially effected by the discussed electronic states at the dielectric interface.

In this section, the influence of different polymer insulators on the OFET charge carrier transport properties has been investigated. The experimental results suggest, that electronic states at the dielectric / semiconductor interface, in the form of Keto or hydroxyl groups, essentially influence the electron charge carrier transport properties in pentacene OFETs. However, at the same time, no influence on the hole charge carrier transport properties could be observed. Therefore the OFET transport properties can be selectively influenced by a modification of the charge carrier trap density at the dielectric / semiconductor interface.

Chapter 5
OFET dielectric interface engineering

In the previous chapter 4 it was demonstrated, that electronic states at the dielectric interface play an essential role with respect to OFET charge carrier transport. The given discussion leads to the question, whether one can modify the organic transistor charge carrier transport properties using dielectric interface engineering by means of charge carrier trap introduction or removal from the dielectric interface. In the current chapter, the effect of Ca traces, deposited on an electron trap afflicted SiO_2 dielectric interface on pentacene electron transport, is investigated in detail. The dependence of the pentacene electron transport on the Ca trace thickness is evaluated and correlated with findings of X-ray photoelectron spectroscopy (XPS) measurements. The experiments, using Ca traces, were conducted to remove charge carrier traps from a SiO_2 dielectric interface. An approach to introduce charge carrier traps to an otherwise inert polymer dielectric is also investigated in detail. The charge carrier traps are introduced to the dielectric interface by exposure to UV radiation in ambient atmosphere, prior to the device completion. The effect of the UV irradiation on the dielectric interface is then characterized, using water contact angle, atomic force and XPS measurements. Finally, the OFETs are characterized electronically, to examine possible effects of the introduced charge carrier traps on the OFET characteristic.

5.1 Ca modified Silicon dioxide

The following experiments are conducted, using the standard pentacene OFET device structure, as described in section 3.2, comprising Ca source-drain electrodes. The investigated devices contain a Ca layer of varying thickness, deposited onto the SiO_2 dielectric. This layer will be referred to as Ca interlayer in the following.

Illustrated in Figure 5.1(a) is the output characteristic of such an OFET comprising a 8Å Ca interlayer. Prior to the illustrated measurement, the device has been exposed to cyclic electrical stress. The purpose of this treatment will be addressed by section 5.1.4. The OFET exhibits exclusive n-type charge carrier transport. The absence of an s-shaped current increase in the linear range of the characteristic indicates a low non-linear contact resistance, as expected for Ca source-drain electrodes, due to a good match of the metal work function (Φ_{Ca} = 2.9eV) and the electron affinity of the organic semiconductor (χ_{Pent} = 2.81eV). An electron mobility of μ_e = $0.167\frac{cm^2}{Vs}$, a threshold voltage of

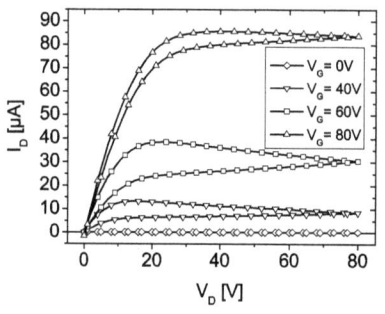 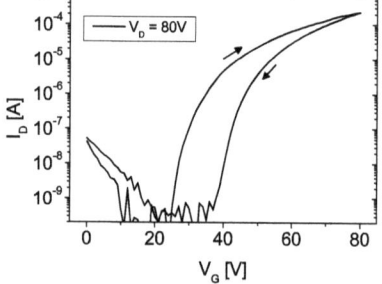

(a) Output characteristic in the electron accumulation.

(b) Transfer characteristic in the electron accumulation. The arrows represent the forward an reverse sweep direction.

Figure 5.1: Output and transfer characteristic of a pentacene OFET incorporating Ca source-drain contacts as well as a SiO_2 insulator comprising a 8Å Ca layer at the interface.

V_{th} = 36 V as well as an $\frac{On}{Off}$ ratio of 10^5 have been derived, using the respective transfer characteristic in the forward sweep direction (See arrows in Figure 5.1(b)). This result stands in clear contrast to the current voltage characteristic of a pentacene OFET without Ca interlayer, but with an otherwise identical device structure. For such a transistor, no measurable electron current could be obtained. The difference in the device performance can be understood as the result of charge carrier traps at the dielectric / semiconductor interface [28], by considering the MIS diode impedance measurements, as discussed in chapter 2.2.2. By taking the previous chapter 4.2 into account, these trap states are proposed to be hydroxyl groups, available at the interface of the SiO_2 dielectric. However, at this stage, the influence of the Ca traces in the investigated device structure is not clear. In the following sections the influence of the Ca interlayer thickness on the pentacene OFET n-type transport will therefore be discussed in detail. Furthermore, the obtained charge carrier transport properties are correlated with findings of X-ray photoelectron spectroscopy measurements, which were performed on standard p^{++} − Si / SiO_2 substrates with varying Ca layer thicknesses.

5.1.1 n-type transport in dependence of a Ca modified SiO_2 interface

In the current section, the n-type transport for pentacene OFETs is investigated in dependence of the Ca interlayer thickness. For this purpose, the parameters μ_e, $V_{th,e}$ as well as the $\frac{On}{Off}$ ratio are extracted from the respective transfer characteristics in the electron accumulation mode. It shall be emphasized, that the investigated OFETs were not subject to cyclic electrical or thermal stress prior to the measurements. Illustrated in Figure 5.2(a) is the electron mobility in dependence of the Ca interlayer thickness. Below thicknesses of 12Å, an enhancement of μ_e is found with a maximum obtained value of μ_e = 0.028$\frac{cm^2}{Vs}$. For thicknesses exceeding 12Å, however, the mobility degrades until μ_e almost vanishes for thicknesses exceeding 26Å. The corresponding threshold voltages and $\frac{On}{Off}$ ratios are shown in Figure 5.2(b). The threshold voltage demonstrates an inverse dependence on the interlayer thickness with an obtained $V_{th,e}$ minimum of 42.2V at a layer thickness of 8Å. The trend of the $\frac{On}{Off}$ ratio graph is similar to the trend of the electron mobility with respect to the Ca passivation thickness. The $\frac{On}{Off}$ ratio development is the consequence of a change in the transistor On-current, due to its dependence on the charge carrier mobility as well the threshold voltage. The Off-current for the investigated devices remains more or less the same. Both the minimum in threshold voltage as well

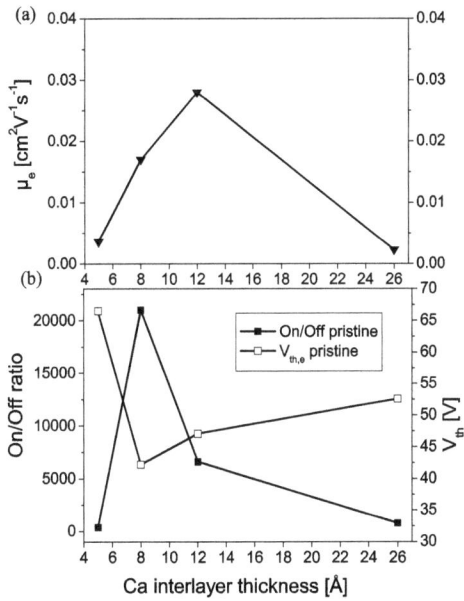

Figure 5.2: a) Pentacene OFET electron mobility as well as b) electron threshold voltage and $\frac{On}{Off}$ ratio, as extracted from the electron accumulation in dependence of the Ca interlayer thickness.

as the maximum in $\frac{On}{Off}$ ratio differs by a Ca interlayer thickness of $\Delta d_{Ca} = 4\text{Å}$ if compared to the interlayer thickness for which the maximum in mobility is obtained.

All of the investigated device parameters indicate an improvement in the OFET charge carrier transport properties up to case dependent Ca interlayer thicknesses of either 8Å or 12Å. This improvement could be the result of doping, a change in charge carrier trap density at the dielectric interface, or possibly due to a change in the semiconductor morphology. However, the experimental fact, that an increase in Ca layer thickness first leads to an improvement and then finally results in a degradation of the electron transport properties is counterintuitive at first. With increasing interlayer thickness, a metallic layer or possibly metallic percolation paths are expected to form on the dielectric surface, short circuiting the source-drain electrodes and therefore reducing the ability to modulate the transistor current by an applied gate voltage. While this effect should negatively influence the OFET charge carrier transport properties, it is not expected to reduce the $\frac{On}{Off}$ ratio by a reduction in the On-current or an increase in the threshold voltage as observed for the conducted experiments. The expected short circuit between the source and drain electrodes was only observed for higher nominal thicknesses of the Ca layer, e.g. for a layer thickness of 250Å. Thus, for the investigated range of the Ca interlayer thickness, no metallic percolation path seems to have evolved between the respective electrodes. In order to investigate, whether the deposition of Ca onto silicon dioxide influences the suggested interfacial trap states by a chemical reaction, X-ray photoelectron spectroscopy measurements were conducted, as discussed in the following section.

5.1.2 XPS interface analysis

The following PES experiments are carried out, using Al Kα radiation, at a sample analyzer angle of 45°. For these experiments, the same standard substrates are utilized as used for the preceding transistor experiments. The substrates consist of p^{++}-Si with a 200nm dry oxide. To conduct the Ca thickness dependent measurements, the metal is stepwise deposited in the DAISY-MAT preparation chamber and XPS measurements are taken, without breaking the ultra high vacuum.

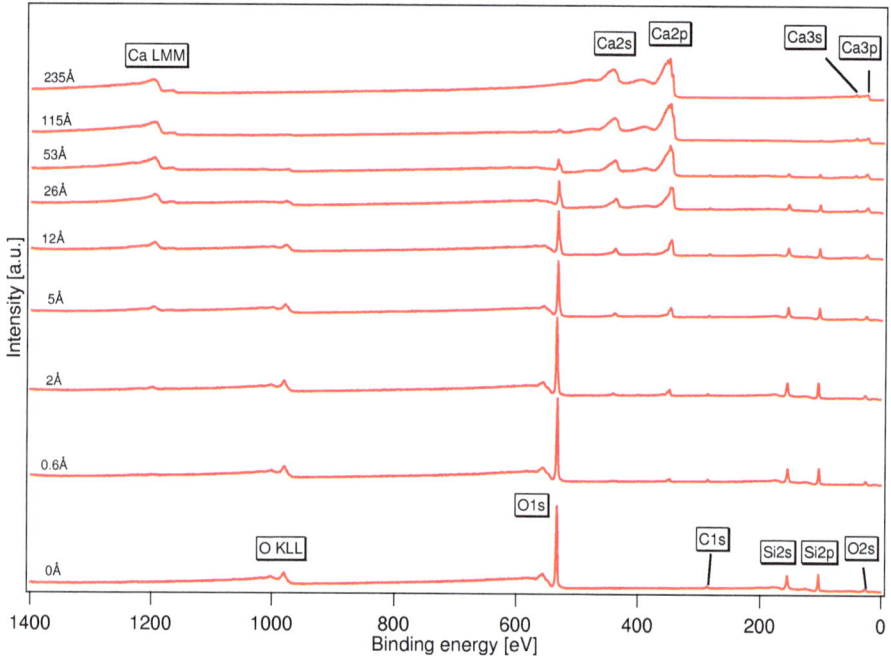

Figure 5.3: XP-survey spectra on SiO$_2$ for ascending Ca layer thickness.

Illustrated in Figure 5.3 are the XP survey spectra, as recorded after the successive Ca deposition steps in ascending order. The formal Ca layer thicknesses were calculated by the use of a calibration measurement, recorded prior to the experiment. As expected, an attenuation of the substrate emission lines (O1s, Si2s and Si2p) is obtained with an increase in Ca adsorbate thickness. However, a closed adsorbate layer is only obtained for Ca layer thicknesses exceeding a formal thickness of 115Å, as indicated by the complete attenuation of the substrate emission lines. This will be further substantiated, by considering the Si2p and O1s emission spectra discussed, later in the text. The obtained C1s emission at a binding energy of 286eV is suggested to be the result of carbohydrate impurities. However, due to its low intensity, it can be concluded, that the substrate cleaning step, as described in section 3.2.2, yields sufficiently clean substrate surfaces.

5.1 Ca modified Silicon dioxide

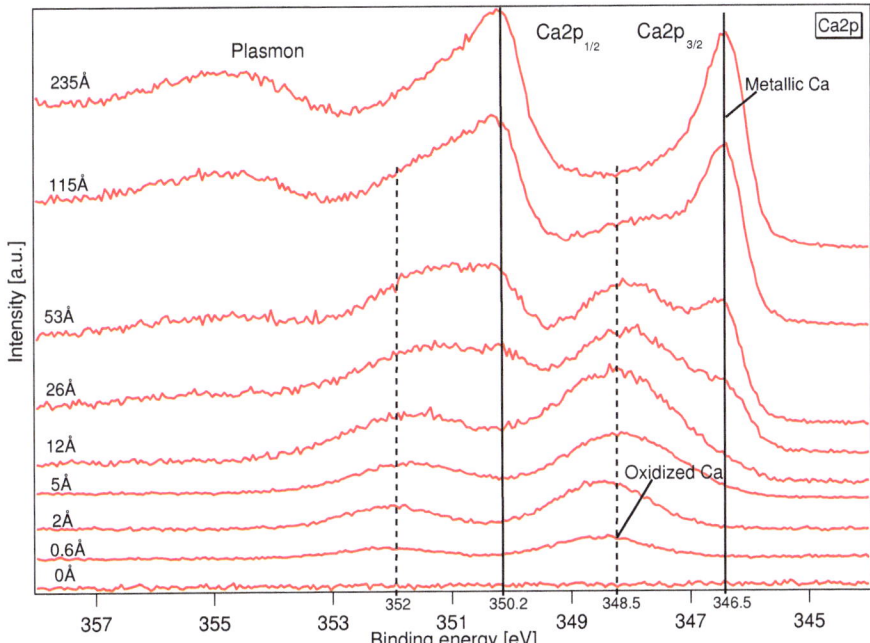

Figure 5.4: Ca2p emission spectra on SiO_2 for ascending Ca layer thickness. The Ca layer thickness is enhanced in between the respective measurements without breaking the vacuum. The spectra have been shifted in binding energy relative to the metallic Ca2p emission at an adsorbate thickness of 235Å for better comparability. The as measured binding energy values are illustrated in Figure 5.7.

Depicted in Figure 5.4 is the Ca2p core level emission spectrum for different Ca layer thicknesses in ascending order. As will be discussed at the end of this section[15], the spectra have been shifted in binding energy relative to the metallic Ca2p emission at an adsorbate thickness of 235Å for better comparability. The as measured spectra are illustrated in Appendix B. For an adsorbate thickness of 0.6Å, $Ca2p_{\frac{3}{2}}$ and $Ca2p_{\frac{1}{2}}$ emission lines were obtained at respective binding energies of 348.5eV and 352eV. With a further increase in thickness, between 5Å and 12Å, the spectrum begins to develop new emission shoulders at 346.5eV and 350.2eV as well as an elevated ground level at high binding energies. For thicknesses exceeding 12Å, the spectrum evolves a plasmon emission at a binding energy of 355eV. Up to a thickness of 53Å, the intensity for all of the detected emission lines increases, in particular a continuous development of the adsorbate components at 346.5eV and 350.2eV is observed. These components assume the typical asymmetric shape of metallic $Ca2p_{\frac{3}{2}}$ and $Ca2p_{\frac{1}{2}}$ emission lines for adsorbate thicknesses exceeding 53Å (Metallic $Ca2p_{\frac{3}{2}}$: 347eV [86]). While the metallic character of the adsorbate is fully developed as of 115Å, a closed metallic layer is only obtained for thicknesses exceeding 235Å. This is indicated by the complete attenuation of the O1s and Si2p emission lines for

---
[15] See Figure 5.7

Surface species Ca2p	Ca		-CaOH		-CaO	
	$2p_{\frac{3}{2}}$	$2p_{\frac{1}{2}}$	$2p_{\frac{3}{2}}$	$2p_{\frac{1}{2}}$	$2p_{\frac{3}{2}}$	$2p_{\frac{1}{2}}$
$\Delta E_{Ca-CaOH}$, ΔE_{Ca-CaO} [eV] [87]	-	-	1.9	2.2	1.3	1.3
Experimental E_{Bind} [eV]	346.5	350.2	348.4	352.4	347.8	351.5
$\Delta E_{2p_{\frac{3}{2}}-2p_{\frac{1}{2}}}$ [eV]	3.7		4.0		3.7	
Fitted E_{Bind} [eV]	-	-	348.7	352.4	347.9	351.6

Table 5.1: Difference in binding energy between Ca and -CaO, -CaOH in a CaNi5 alloy. Resulting Ca2p -CaO and -CaOH binding energies for Ca on SiO$_2$.

Ca interlayer thicknesses of 235Å, as depicted in Figures 5.6(a) and 5.6(c). By considering the Ca2p emission spectrum (Figure 5.4) for layer thicknesses below 115Å, however, the adsorbate exhibits, in addition to the metallic component, an oxidized Ca component as identified for binding energy values of 352eV and 348.5eV [76]. With a declining layer thickness, the metallic Ca fraction is continuously reduced until, for layer thicknesses below 12Å, the adsorbate is even reasoned to be mainly composed of oxidized Ca, possibly containing a small metallic fraction. This is concluded from the first indication of the metallic Ca fraction between adsorbate thicknesses of 5Å and 12Å, by the development of the described metallic Ca emission line shoulders at 350.2eV and 346.5eV as well as an elevated ground level at higher binding energies.

A detailed analysis of the Ca2p emission spectra shows, that the oxidized Ca component in the adsorbate contains calcium oxide (-CaO) as well as Ca hydroxide (-CaOH). The individual binding energies of the oxide components can be approximated by adding the difference in binding energies for -CaO and -CaOH, as published with regard to the metallic Ca emission of CaNi5 alloys [87], to the experimentally obtained metallic Ca$2p_{\frac{3}{2}}$ and Ca$2p_{\frac{1}{2}}$ energies. The difference in energy for -CaOH ($\Delta E_{Ca-CaOH}$) and -CaO (ΔE_{Ca-CaO}) as well as the resulting binding energy values E_{Bind} for the Ca hydroxide and oxide components are listed in Table 5.1. A fit using the respective oxidized components is given in Figure 5.5. Here, the oxidized Ca components are fitted to the Ca2p emission spectrum for an adsorbate thickness of $d_{Ca} \approx 53$Å. In order to obtain a correct fit, the metallic Ca fraction was subtracted, using the scaled metallic Ca2p emission recorded for an adsorbate thickness of 235Å[16]. The fit is implemented, using a Voigt function, under the consideration of an intensity ratio of 1:2

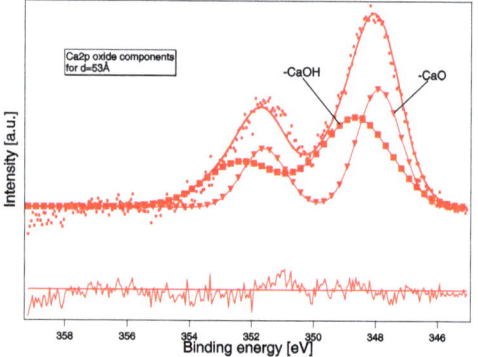

Figure 5.5: Ca2p emission spectrum of only the oxidized component for an adsorbate thickness of d=53Å. The metallic component has been subtracted using the scaled Ca2p emission at d=235Å. The fit exhibiting the closed triangle plot represents the -CaO component, while the closed square plot represents the -CaOH component. The residuum of the fit is given at the bottom of the graph.

---

[16] The scaling was conducted by the use of an available IGOR macro.

5.1 Ca modified Silicon dioxide

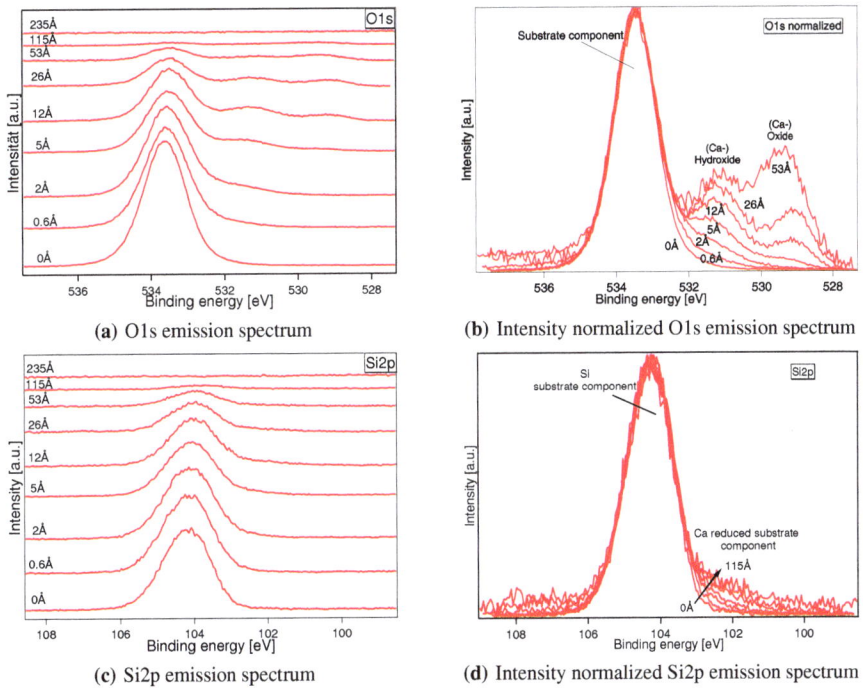

Figure 5.6: O1s and Si2p emission spectra. The spectra have been shifted in binding energy with respect to the substrate emission line obtained for a Ca thickness of 0Å, for better comparability. The as measured binding energy values are illustrated in Figure 5.7. The as measured spectra are illustrated in Appendix B.

for the respective Ca2p$_{\frac{1}{2}}$ and Ca2p$_{\frac{3}{2}}$ dublett states. The split in binding energy for these states is assumed to be constant for both oxide components ($\Delta E_{2p\frac{3}{2}-2p\frac{1}{2}} \approx 3.7$). For the fit only, the intensity ratio as well as the split in binding energies were held constant. All other parameters, such as the binding energy, the full width half maxiumum of the peaks as well as the intensity values were varied. The residuum at the bottom of Figure 5.5 only indicates a small deviation between the experimental data and the conducted fit. The respective -CaOH and -CaO binding energy values obtained by the fit are summarized in Table 5.1. Indeed, a good match of these values to the oxidized Ca binding energies is obtained by considering the published data of Selvam et al.

The occurrence of -CaO as well as -CaOH in the Ca adsorbate is further substantiated by considering the intensity normalized O1s emission spectrum as illustrated in Figure 5.6(b). This spectrum is depicted for adsorbate thicknesses between 0Å and 53Å. The binding energy values of the obtained spectra have been corrected with respect to the oxygen substrate component (533.3eV) for better comparability. In addition to the oxygen substrate component, two further components at binding energy values of 529.2eV and 531.55eV are found, which are enhanced with increasing adsorbate thickness. These components are identified as Ca oxide and Ca hydroxide, using literature values

(-CaOH O1s: 531-532eV [76], -CaO O1s: 529.9eV [86]). Due to the absence of atmospheric oxygen in the UHV (base pressure $\approx 10^{-10}$ mbar) of the DAISY-MAT preparation and measurement chambers, the formation of -CaO and -CaOH for low adsorbate thicknesses can only be the result of an interface reaction between Ca and the SiO_2 oxygen components as well as available hydroxyl groups. The intensity normalized Si2p emission spectrum, depicted in Figure 5.6(d), supports this suggestion. The binding energy values of the individual spectra have been corrected with respect to the Si (103.98eV) substrate component. Upon Ca deposition, an emission shoulder at ≈ 102eV develops in addition to the Si substrate component. By considering the Si2p emission line to be subject to a chemical shift of ≈ 1eV per oxidation state [76], the observed shoulder is ascribed to Si^{2+} in the reduced SiO_2 as the result of the implied interface reaction with Ca.

The emission spectra of Figures 5.4, 5.6(a), 5.6(c) as well as the intensity normalized Si2p and O1s emission spectra illustrated in Figures 5.6(d) and 5.6(b) have been shifted in binding energy for better comparability. The Ca2p emission spectrum has been shifted with respect to the 235Å metallic Ca2p emission at $E_{Bind} = 344.94$eV, while the Si2p and O1s emission spectra have been shifted with respect to their individual substrate emission lines at $E_{Bind} = 103.98$eV and 533.3eV. The as measured binding energy values of the respective emission spectra are illustrated in Figure 5.7. A ΔE_{Bind} of 245eV and -183.14eV has been added to the respective $Si2p_{\frac{3}{2}}$ and O1s binding energy values, in order to be able to visualize the binding energy trend for the Ca, oxygen and silicon components in the same Figure.

Up to an adsorbate thickness of 26Å, the oxidized $Ca2p_{\frac{3}{2}}$ adsorbate component shifts to lower binding energies by a value of ≈ 1.9eV. The observed shift is then reversed with increasing adsorbate thickness by a value of ≈ 1eV, which seems to be correlated with the appearance of the distinct metallic $Ca2p_{\frac{3}{2}}$ emission. By considering the $Si2p_{\frac{3}{2}}$ as well as the O1s substrate components, it is found, that the shift to lower binding energies for an increase in Ca layer thickness up to $d_{Ca} = 26$Å is significantly less than the shift observed for the oxidized Ca component. However, the metallic $Ca2p_{\frac{3}{2}}$, the $Si2p_{\frac{3}{2}}$ as well as the O1s substrate components shift to higher binding energies by $\Delta E_{Bind} \approx 1.5$eV for $d_{Ca} > 26$Å. While the final cause for the observed shifts is not clarified, they are likely to be the result of charge transfer effects between the substrate and the adsorbate, due to the use of an insulating substrate (SiO_2). It is therefore proposed, that the obtained shift to lower binding energies of the oxidized Ca layer, up to a thickness of 26Å, is due to a negative charging of the adsorbate by substrate primary and secondary photoelectrons. The subsequent shift to higher binding energies for all

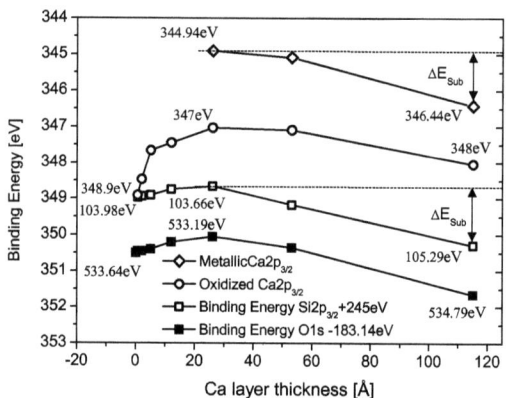

Figure 5.7: As measured binding energies for the Ca2p, Si2p and O1s emission lines, in dependence of the Ca adsorbate thickness.

of the adsorbate and substrate components is suggested to be the result of a final alignment in Fermi level between the evolving metallic Ca component and the substrate. This leads to a change in electric field across the oxidized Ca layer and therefore explains the observed loss in photoelectron kinetic energy.

5.1.3 Correlation between transistor performance and PES data

With respect to the investigated OFET electron transport parameters in dependence of the Ca interlayer thickness, as depicted in Figure 5.2, it was found, that the electrical transistor parameters improve up to a layer thickness of $\approx 12\text{Å}$. For interlayer thicknesses exceeding this value, however, the charge carrier transport properties are significantly degraded, until they are almost completely impeded for thickness values exceeding $\approx 26\text{Å}$. By considering the XPS measurements, it is found, that the deposition of Ca on SiO_2 leads to the formation of an oxidized Ca layer, containing -CaOH and -CaO for low layer thicknesses. This formation of an oxidized Ca layer on top of the SiO_2 correlates well with the improvement in electron charge carrier transport properties up to a Ca interlayer thickness of $\approx 12\text{Å}$. Hence, it is suggested, that this additional dielectric layer isolates and compensates available electron traps at the dielectric SiO_2 interface, allowing for the realization of n-type pentacene OFETs. Since -CaOH is known for its ionic binding, it is likely, that Ca atoms donate electrons and thereby neutralize the available hydroxyl groups on the SiO_2 interface, which are suggested to serve as electron traps by Chua et al. [29]. For Ca adsorbate thicknesses exceeding 12Å, the degradation of the OFET electron transport properties correlates well with the first indication of a metallic fraction in the Ca2p emission spectrum. An almost complete attenuation in the OFET performance is found for Ca layer thicknesses above 26Å, in accordance with the development of the metallic $Ca2p_{\frac{3}{2}}$ and $Ca2p_{\frac{1}{2}}$ emission lines. Therefore, it seems safe to conclude, that the increasing presence of a metallic Ca fraction is responsible for the observed loss in the OFET performance with increasing Ca passivation layer thickness. A possible explanation of the underlying physics would be, that the metallic fraction partially screens the electric gate field in the transistor channel and therefore negatively affects the device parameters.

5.1.4 Influence of thermal and electrical stress on OFET transport properties

In the previous sections it was demonstrated, that n-type pentacene OFETs can be realized on SiO_2 dielectrics, modified by the deposition of a thin Ca layer. Utilizing X-ray photoelectron spectroscopy, it was substantiated, that this type of charge carrier transport is due to the formation of an additional Ca oxide dielectric on top of the SiO_2 insulator for low Ca layer thicknesses. This additional dielectric, containing the oxide components -CaOH and -CaO, was suggested to cover or eliminate electron traps available at the SiO_2 interface. However, it was also observed that the n-type charge carrier transport properties are degraded for Ca layer thicknesses exceeding 8Å-12Å. This degradation was ascribed to the formation of a metallic component in the oxidized Ca layer. In the following, the influence of thermal and electrical stress with respect to the constitution of the Ca layer as well as the influence of the applied stress om OFET charge carrier transport properties is discussed.

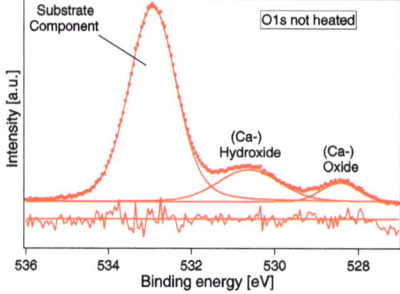
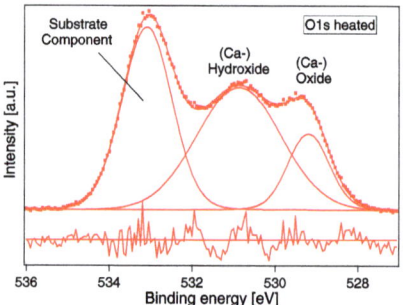

(a) O1s emission spectrum of an unheated substrate.

(b) O1s emission spectrum of a substrate exposed to T=180°C for the duration of t=1h.

Figure 5.8: O1s emission spectra for an unheated / heated Si p^{++}-Si substrate with a 200nm thermally grown dry oxide comprising a Ca adsorbate (d_{Ca} = 50Å). The temper step was conducted in the UHV. Fits to the individual components of the spectra as well as their error function at the bottom of each spectrum are illustrated by both Figures. The 5 times magnified residuum indicates the deviation of the fit to the measurement.

For these experiments, a standard substrate ($p^{++} - Si / SiO_2$) is covered with a 50Å thick Ca layer in the preparation chamber of the DAISY-MAT and subsequently annealed at $T = 180°C$ for the duration of t = 1h, without breaking the vacuum. The PES investigation was conducted, using Al Kα radiation at a sample analyzer angle of 45°. The findings of these measurements prior to and after the substrate annealing are correlated with the transport properties obtained for OFETs, exposed to either thermal stress during their production process or cyclic electrical stress. The investigated OFETs are realized using the standard OFET device structure, as discussed in section 3.2, comprising a Ca interlayer of varying thickness (d_{Ca} = 0Å-26Å) between the SiO_2 and pentacene layers. The source-drain metalization is chosen as either Ca or Au.

XPS annealing experiment

Illustrated in Figure 5.8(a) is the O1s emission spectrum of the $p^{++} - Si/SiO_2$ substrate, comprising a d_{Ca} = 50Å thick Ca layer. The spectrum was obtained prior to the temperature treatment of the sample. As already elaborated in section 5.1.2, the spectrum yields, in addition to the O1s substrate component at a binding energy value of 532.94eV, the oxidized Ca components -CaOH and -CaO at respective binding energies of 530.63eV and 528.45eV. Depicted in Figure 5.8(b) is the O1s emission spectrum of the same sample in its annealed state. A pronounced relative intensity increase for the oxidized Ca components with regard to the oxygen substrate component is obtained, if compared to the respective oxygen components, as discussed for the unheated sample. According to the fits presented in Figures 5.8(a) and 5.8(b), the area fraction of the -CaOH component is enhanced from a value of 16.18% up to 45.24%, and the area fraction of the -CaO component is enhanced from 5.57% up to a value of 14.23%. This approximation is valid under the assumption that the total amount of oxygen, available in the substrate, remains constant during the annealing step. The utilized fit parameters for the applied

5.1 Ca modified Silicon dioxide

O1s component	Substrate unheated			Substrate annealed		
	BE [eV]	FWHM [eV]	area fraction[%]	BE [eV]	FWHM [eV]	area fraction[%]
Substrate Component	532.94	1.16	78.25	533.1	1.37	40.53
-COH	530.63	1.55	16.18	530.87	2.38	45.24
-CO	528.45	1.05	5.57	529.2	1.11	14.23

Table 5.2: Oxygen components for a p^{++} – Si / SiO$_2$(200nm) substrate comprising a d_{Ca} = 8Å Ca layer in its unheated and annealed state. The substrate was annealed at T = 180°C for the duration of t = 1h.

Voigt function, after the subtraction of the inelastic background, using a Shirley function [81], as well as the individual area fractions are summarized in Table 5.2. The residuum at the bottom of each Figure represents the deviation of the fits with respect to the measurement. This experimental result implies, that the discussed Ca oxidation reaction for thin films deposited onto SiO$_2$ is promoted by heat, reducing the metallic content in the Ca layer.

OFET annealing experiments

In the previous section was shown, that a metallic fraction in the Ca interlayer has a negative effect on the charge carrier transport. In the following it is therefore investigated, whether the heat induced enhancement of the oxidation reaction has any effect on the transistor operation. For this experiment, the standard OFET architecture, using a p^{++} – Si / SiO$_2$(200nm) substrate passivated by a 8Å thick Ca layer, was utilized. However, after the pentacene deposition and prior to the deposition of the Ca source-drain contacts, the substrates were exposed to an annealing step on a hot plate in inert N$_2$ atmosphere for the duration of 1h at different temperatures. For the annealing, temperatures of 100°C, 140°C and 160°C were applied.

The influence of the different annealing steps with regard to the OFET electron charge carrier mobility and threshold voltage is illustrated in Figure 5.9. The demonstrated values were extracted, using the respective transfer characteristics in the electron accumulation mode at room temperature. Up to an annealing tem-

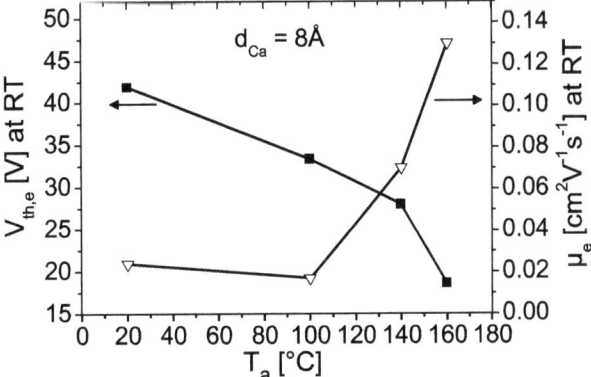

Figure 5.9: Electron mobility and threshold voltage (at RT) for OFETs comprising a d_{Ca} = 8Å Ca passivation layer, in dependence of different annealing temperatures during the production process. The annealing was conducted in inert N$_2$ atmosphere for the duration of t=1h.

Temper State	$\mu_h [\frac{cm^2}{Vs}]$	$V_{th,h}$ [V]	$\frac{On}{Off}$
Pristine	0.2	-0.3	$\approx 10^4$
Tempered@160°C	0.1	7.2	$\approx 10^3$

Table 5.3: p-type pentacene OFET charge carrier transport parameters at RT. The values were extracted prior to and after an annealing step at T=160°C for the duration of t=1h.

perature of $T_a = 100°C$, the electron mobility remains almost constant, while $V_{th,e}$ is already significantly reduced from 42 V to 33 V. A further increase in the annealing temperature, up to a value of $T_a = 160°C$, leads to a drastic decrease for $V_{th,e}$ (19 V) as well as to a steep increase in the electron mobility μ_e, up to a value of $0.13 \frac{cm^2}{Vs}$. The $\frac{On}{Off}$ ratio dependence evolves in accordance with the electron mobility. This parameter is improved from $7.5 * 10^3$ for the pristine device, up to a value of $8.8 * 10^5$ for the OFET exposed to an annealing temperature of $T_a = 160°C$.

The observed improvement in device performance is most probable due to the demonstrated enhanced Ca oxidation reaction or heat induced morphology changes. To rule out a parameter improvement, due to morphology changes of the organic semiconductor during the substrate annealing, a temper experiment, using a pentacene p-type OFET with an Au source-drain metalization and a pristine SiO_2 / pentacene interface was conducted. For this experiment, the OFET was characterized in its pristine state and subsequently annealed on a hotplate at $T_a = 160°C$ in N_2 atmosphere for the duration of 1h. The successive characterization was conducted at room temperature. Even though this experiment yields only information on p-type charge carrier transport, heat induced morphological changes should influence hole charge carrier transport properties comparable to those of electrons. This is implied, since a change in morphological order would similarly influence the overlap in π-orbitals, contributing to the HOMO as well as the LUMO transport levels, as discussed in section 2.1. The extracted device parameters, prior to and after the conducted annealing step, are summarized in Table 5.3. Due to the annealing of the OFET, the charge carrier mobility is reduced by a factor of 2, while the $\frac{On}{Off}$ ratio is reduced by one order of magnitude. Taking into account the observed negative impact of the heat treatment at $T_a = 160°C$ on the hole transport properties, it is suggested, that the morphological order of the semiconductor layer, and therefore the $\pi - \pi$ overlap in the transport direction, is degraded by the annealing.

Considering the discussed results, it is conjectured, that the observed improvement in the n-type charge carrier transport for annealed OFETs, comprising a Ca interlayer, is not dominated by morphology changes, but is rather a result of the discussed heat induced Ca oxidation reaction. Furthermore it is implied, that possible changes in the semiconductor morphology, as well as the observed enhancement in oxidation reaction of the Ca passivation layer, are competitive processes with respect to the OFET charge carrier transport. This suggests, that the n-type charge carrier transport properties could be further enhanced by an improvement in the Ca oxidation reaction, without the negative influence of a degraded pentacene morphology.

5.1 Ca modified Silicon dioxide

OFETs exposed to cyclic electrical stress

It was demonstrated by Ahles et al. [28], that the n-type pentacene OFET performance for devices comprising an oxidized Ca interlayer, as well as a Ca source-drain metalization, can be significantly improved by exposing the transistors to cyclic electrical stress in a nitrogen atmosphere. For such a device conditioning, the gate voltage V_G is pulsed between 0 and 80V in 5s intervals for the duration of 1h, while a constant drain voltage of $V_D = 80V$ is applied. Depicted in Figure 5.10 is the improvement in drain current for an n-type OFET, comprising a 8Å Ca interlayer, during the electrical cyclic conditioning step. A continuous improvement in drain current from $\approx 15\mu A$ to $\approx 65\mu A$,

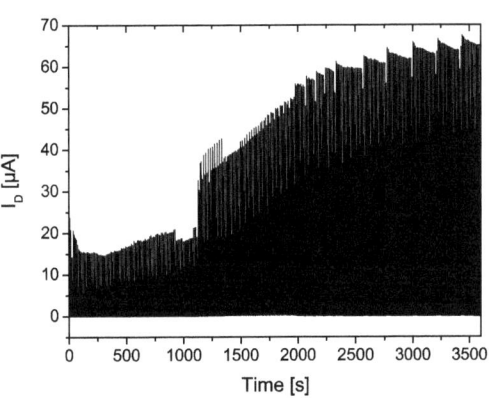

Figure 5.10: Improvement in I_D for an n-type OFET comprising a 8Å passsivated SiO$_2$ dielectric, during the described electrical cyclic conditioning step.

with an indicated saturation of the drain current at the end of the cyclic conditioning, is obtained. The irreversible improvement in device performance is not limited to the transistor On-current, but also positively affects such parameters as electron mobility, threshold voltage and $\frac{On}{Off}$ ratio. The improved device parameters are summarized in Table 5.4, in comparison to the parameters obtained for an unconditioned transistor and an annealed OFET, both comprising a 8Å Ca passivated SiO$_2$. By comparing the electrical cyclic conditioned OFET to the unconditioned device, an improvement in electron mobility as well as the $\frac{On}{Off}$ ratio by approximately one order of magnitude is obtained, while the electron threshold voltage is reduced by a value of about 6V. Furthermore it can be seen, that a transistor, exposed to the cyclic electrical stress, exhibits a comparable improvement in OFET device parameters as obtained for a transistor exposed to the discussed annealing step at a temperature of T = 160°C. However, as indicated in Table 5.4, the threshold voltages for electrical conditioned and annealed OFETs differ in the obtained minimum value by $\Delta V_{th,e} \approx 19V$.

While the origin of the heat induced improvement in OFET parameters for Ca passivated devices seems to be clarified, the reason for the performance improvement of OFETs, conditioned by the

Conditioning	$\mu_e [\frac{cm^2}{Vs}]$	$V_{th,e}$ [V]	$\frac{On}{Off}$
Pristine	0.017	42.2	$\approx 10^4$
El. Cyclic Cond.	0.167	36.0	$\approx 10^5$
Tempered @ 160°C	0.14	17.0	$\approx 10^5$

Table 5.4: Comparison in OFET charge carrier transport parameters for devices containing a 8Å Ca passivation layer in their pristine, electrical cyclic conditioned and annealed state.

cyclic electrical stress, still requires discussion. Illustrated in Figure 5.11 are the output characteristics for a pristine and an annealed (1h at $T_a = 160°C$) OFET. Both transistors contain an 8Å thick Ca interlayer. As expected, the current voltage characteristic of the pristine OFET could be significantly improved by the electrical cyclic conditioning. As already indicated by the device parameters summarized in Table 5.4, the improved drain current for the OFET, exposed to electrical stress, is comparable to the improved current obtained for the annealed OFET. However, an additional electrical conditioning of the already thermally annealed device does not yield a further significant device improvement. It is therefore suggested, that both the electric cyclic conditioning as well as the annealing rely on the same device changes, namely an enhanced oxidation reaction of the Ca interlayer.

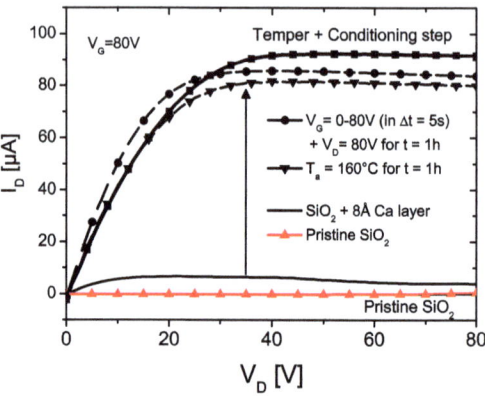

Figure 5.11: Comparison of the output characteristics for pentacene OFETs comprising a 8Å Ca interlayer in their pristine, annealed ($t = 1h$ at $T_a = 160°C$) and electrically cyclic conditioned state. Furthermore the characteristic for an OFET comprising a pristine SiO_2 / pentacene interface is illustrated.

The improvement of the OFET performance, in dependence of the Ca layer thickness, is summarized in Figures 5.12(a) and 5.12(b). Depicted in Figure 5.12(a) is the development of the electron mobility for OFETs in their prisitine and their electrically cyclic conditioned state. Illustrated in Figure 5.12(b) are the improved threshold voltage and $\frac{On}{Off}$ ratio values. The improvement in OFET device performance, due to the enhanced oxidation reaction, is discriminative for different Ca layer thicknesses. For thicknesses below ≈ 12Å, the device parameters improve stronger with increasing Ca layer thickness. This means, that the SiO_2 surface is covered more and more with oxidized Ca, thereby reducing the impact of the electron traps on the device performance. On the other hand, the metallic fraction in the layer is also increased, which negatively influences the device improvement. Since in this thickness range the metallic Ca fraction of the layer may still be small enough to be fully oxidized during the promoted oxidation reaction, the positive influence of the Ca passivation prevails. For thicknesses exceeding 12 Å, however, the influence of the metallic Ca becomes strongly visible. In this range, the metallic Ca fraction is no longer sufficiently oxidized by the cyclic electrical or heat treatment steps, leading to a substantial metallic fraction in the CaO-layer. Therefore, a further increase in Ca layer thicknesses does not result in the desired improvement.

In the previous section it was demonstrated, that n-type charge carrier transport can be realized for pentacene OFETs comprising a SiO_2 dielectric, as long as the insulator surface is modified by the deposition of Ca traces. This result is in accordance with data published by Ahles et al. [28]. Furthermore, it was substantiated, that the n-type OFET performance is strongly dependent on the Ca interlayer thickness. An increase in the OFET performance is obtained up to thickness values

5.2 UV modified Polymethylmetacrylate

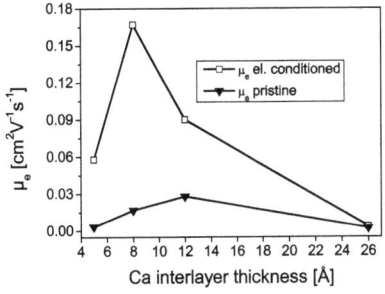
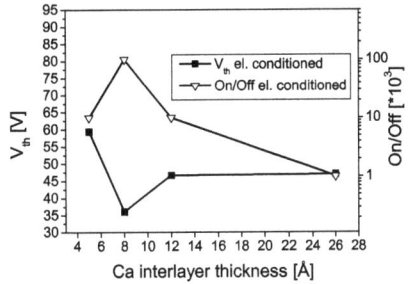

(a) Electron mobility for OFETs in their pristine and electrical cyclic conditioned state.

(b) $V_{th,e}$ and $\frac{On}{Off}$ ratio for OFETs in their electrical cyclic conditioned state.

Figure 5.12: OFET charge carrier transport parameters in dependence of the Ca passivation thickness.

between d_{Ca} = 8 - 12Å. However, for thicknesses exceeding this range, a degradation in the OFET n-type performance is found, until it is almost completely inhibited for Ca trace thicknesses above d_{Ca} = 26Å. The improvement in transistor performance could be linked to the formation of an oxidized Ca layer on top of the SiO$_2$ insulator, and thus isolating or compensating available charge carrier traps at the dielectric surface. The obtained degradation in n-type performance for a Ca layer thickness exceeding d_{Ca} = 12Å was concluded to be the result of the formation of a metallic Ca fraction in the oxidized Ca layer. The almost completely inhibited n-type performance for Ca layer thicknesses exceeding d_{Ca} = 26Å is therefore conjectured to result from the metallic character of the interlayer. Finally it was demonstrated, that the n-type performance of such OFET devices can be significantly improved, if the oxidation reaction of the deposited Ca layer is enhanced. This can be achieved by exposing the OFETs to thermal stress during their production process, or by applying a cyclic electrical conditioning step to the completed devices.

5.2 UV modified Polymethylmetacrylate

In chapter 4, the influence of trap states at the dielectric interface with regard to OFET charge carrier transport properties was discussed. Furthermore, in the first section of chapter 5, a possibility was investigated by which such electronic states, available on SiO$_2$ dielectric interfaces in the form of hydroxyl groups, can be covered or annihilated by using traces of Ca. In the current section it will be demonstrated, that the introduction of charge carrier traps to polymer dielectric interfaces can be used to change the OFET polarity from unipolar n-type to unipolar p-type.

5.2.1 Introduction of charge carrier traps on PMMA

In order to introduce electronic trap states to a PMMA interface, the insulators are irradiated, using UV radiation in ambient atmosphere. The UV irradiation of PMMA results in the formation of keto

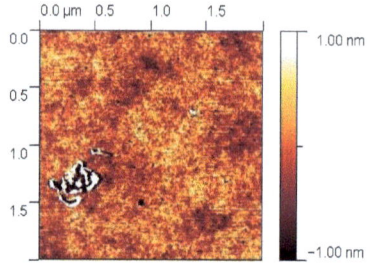 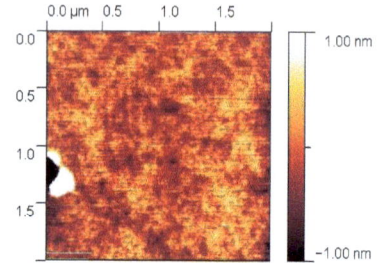

(a) Topography of a PMMA layer in its pristine state.

(b) Topography of a PMMA layer UV irradiated in ambient atmosphere for $t = 10$min.

Figure 5.13: Comparison in PMMA topography for a pristine and UV irradiated layer deposited on SiO_2.

and hydroxyl groups in the polymer near surface layer [61, 63, 88]. Such functional groups are suggested to negatively influence or even completely inhibit the n-type pentacene OFET charge carrier transport, as described in chapter 4.2. However, prior to considering the impact of UV irradiated PMMA dielectrics on the pentacene OFET performance, the interface properties and the chemical composition of pristine and UV modified thin films are investigated, using water contact and XPS measurements. Furthermore, possible changes in the pentacene topography for thin films deposited onto pristine and UV modified PMMA substrates are examined by using atomic force microscopy.

Interface analysis

The samples, which are investigated in the following, are prepared, using the standard p^{++} – Si / SiO_2 substrates, on top of which a $\approx$ 119nm PMMA layer is deposited by spin coating, using a 2%wt THF solution. The exact process parameters / dilutions are listed in Appendix D. Selected samples are then irradiated by UV radiation in ambient atmosphere for a time period between 2 and 15 minutes, using wavelengths of 254 and 185nm with a respective intensity of $15 \frac{mW}{cm^2}$ and $1.5 \frac{mW}{cm^2}$. To stabilize the radiation intensity, the utilized ozone reactor is preheated prior to the exposure for a time frame of 10 minutes.

Illustrated in Figures 5.13(a) and 5.13(b) are the respective AFM measurements of a pristine and a UV modified PMMA layer, recorded in air. The UV exposure was conducted for a time frame of $t = 10$ minutes. The micrographs were obtained in the non contact mode and measured for an area of 2 x 2μm^2. The surface roughness of the investigated samples was derived, excluding large defects, which are most likely caused by substrate impurities. For both samples a rms roughness value of $R_q < 1$nm was obtained. The UV modified sample exhibited a roughness of $R_q \approx 0.26$nm, which is lower than the value obtained for the pristine sample with $R_q \approx 0.36$nm. However, since the vertical resolution of an AFM, operated in air, is limited to $\approx$ 1nm, due to adsorbates such as water, it is concluded, that the surface roughness remains unchanged during the UV exposure.

5.2 UV modified Polymethylmetacrylate

While the surface roughness of the PMMA thin films stays below 1nm during the UV exposure, the water contact angle (CA) is significantly affected. Depicted in Figure 5.14 is the water contact angle for PMMA thin films in dependence of their UV exposure time. For its pristine state, the PMMA layer exhibits a contact angle of 80°, which is degraded during the investigated irradiation time down to a value of 30°. The largest degradation in contact angle seems to occur during the first 5 minutes of the exposure, where the contact angle is degraded by 40°. A further increase in exposure time by 10 minutes leads to an additional degradation in the CA of only 10°. This indicates, that the interface reactions, leading to the degradation in CA, are more or less finalized during the first 10 minutes of the UV exposure. The obtained degradation in CA is interpreted as an increase in oxygen containing polar groups in the near surface layer of the polymer, due to photooxidation [59]. This is suggested, since the surface roughness stays below 1nm during the investigation, and the experimental parameters as well as the ambient conditions remained the same for the respective measurements. The photooxidation is assumed to be the result of different reaction mechanisms between oxygen and molecular radicals.

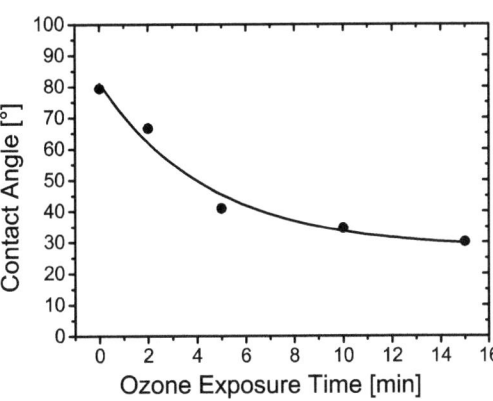

Figure 5.14: Water contact angle for PMMA in dependence of the UV exposure time.

To further quantify the photooxidation process during the UV exposure, oxygen containing components, available at the surface of pristine and irradiated PMMA dielectrics (10 minutes), were investigated, using PES. The measurements were conducted, utilizing MgKα radiation at a sample analyzer angle of 45°. The standard p^{++} – Si / SiO$_2$ substrates were employed, in order not to influence the film forming properties of PMMA. To reduce substrate charging effects during the PES investigation, a reduced dielectric thickness of 100nm was chosen.

The result of the PES investigation is illustrated in Figures 5.15(a) and 5.15(b) in the form of an area normalized O1s and C1s emission spectrum for PMMA in its pristine (straight line) and UV modified state (filled square scatter plot). For better comparability, all of the spectra have been normalized to an area of one after the removal of the inelastic background, using a Shirley function [81].

By first considering the O1s emission obtained for the pristine PMMA sample, it is clearly visible, that the emission line is composed of two components of almost equal intensity. Due to the UV irradiation step, however, the relative intensity of the oxygen component at lower binding energies is significantly increased, leading to a change in the form of the O1s emission as well as a shift of its maximum to lower binding energies. The shape of the C1s emission spectra, depicted in Figure 5.15(b), indicates the presence of several carbon components in the chemical structure of the investigated PMMA thin films. Due to the UV exposure, the relative intensity of the carbon component at

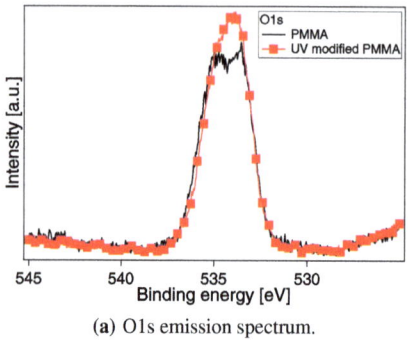 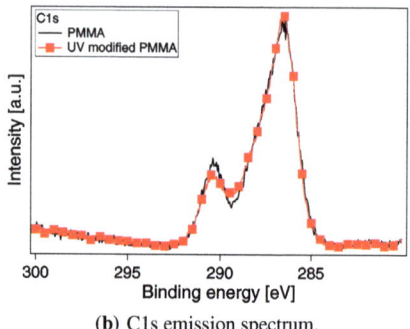

(a) O1s emission spectrum.　　(b) C1s emission spectrum.

Figure 5.15: Area normalized O1s and C1s emission spectra of PMMA and UV modified PMMA.

291eV is decreased, while a relative increase in intensity of the C1s emission at a binding energy of 289eV is observed.

By considering the PMMA chemical structure, illustrated in Figure 5.16, as well as data published by others (O=C O1s: 532.2eV; O-C O1s: 533.7eV [59, 61]), the oxygen components indicated by the O1s emission can be ascribed to O=C and O-C components at respective binding energies of 533.6eV - 533.75eV (a) and 535.15eV - 535.2eV (b). These values have been extracted, using fits to the respective O1s emission lines of the pristine and irradiated sample, as illustrated in Figures 5.17(a) and 5.17(b). The fits have been obtained, utilizing a Voigt function, and show a very good agreement with the experimental data. The residuum illustrated at the bottom of each figure indicates the deviation of the fit from the experimental data. All of the parameters extracted from these fits are summarized in Table 5.5. The observed difference in binding energy of ≈ 1.2eV - 1.5eV, with respect to the oxygen component binding energy values published by Hozumi et al. [59], is ascribed to a positive charging of the substrate during the measurement.

As expected for the pristine sample, due to the PMMA chemical composition, the area fraction of the individual oxygen components is almost equal at 45.3% (O-C) and 44.6% (O=C) with a difference, however, to the theoretical value of 50% by ≈ 5% for both oxygen components. This discrepancy is suggested to be the result of inaccuracies in the fit, due to the inelastic background removal using the Shirley method [81].

The deconvolution of the O1s emission spectrum, obtained for the UV modified sample, is conducted by assuming, that the line width used for the O1s core level components of the pristine sample remains unchanged. For this fit, only the intensity and binding energy values are varied. The fit yields a pronounced relative increase in area fraction of the O=C component by 7.89%, while the area fraction of the O-C component is reduced by 5.68%

Figure 5.16: PMMA chemical structure O=C is marked by a), and O-C is marked by b)

5.2 UV modified Polymethylmetacrylate

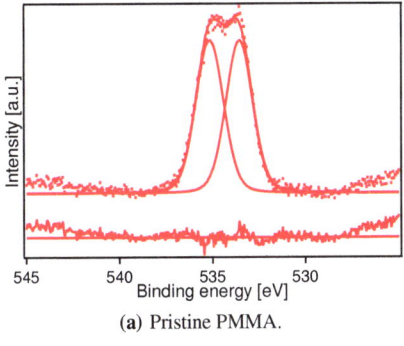
(a) Pristine PMMA.

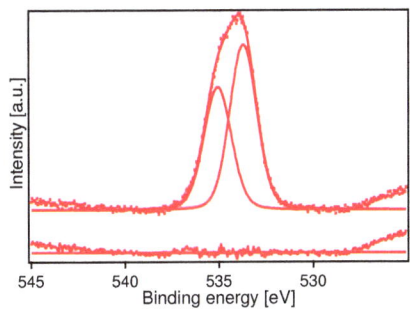
(b) UV irradiated PMMA (in ambient atmosphere for t=10min.).

Figure 5.17: O1s PMMA emission lines, recorded using Mg Kα radiation at a sample / analyzer angle of 45°. The error graph at the bottom of each figure indicates the deviation between experimental data and the conducted fit.

Taking the intensity of the respective oxygen and carbon emissions for the pristine and UV modified samples into account, as well as the atomic sensitivity factors $S_O = 0.711$ and $S_C = 0.296$ for oxygen and carbon [76], an increase in the substrate oxygen concentration by ≈0.76% is obtained, due to the UV exposure, while the carbon concentration is reduced by the same amount. These values are derived, using equation 3.4, under the assumption, that the change in PMMA stochiometry is homogeneous for the entire sample interface. The obtained result is in line with work conducted by Hozumi et al. [63], who have performed a similar PES experiment on pristine and UV modified PMMA samples. Even though the increase in oxygen concentration is small, it appears, that oxygen is incorporated into the PMMA layer during the UV exposure by photooxidation. As a result of the pronounced relative increase in Keto groups, an incorporation of oxygen into the PMMA layer is most likley, due to an absolute increase in O=C groups.

So far, the discussion implies an increase in oxygen containing polar groups in the near surface layer of UV irradiated PMMA insulators, due to an increase in keto groups. However, the expected

O1s components	PMMA pristine			PMMA UV modified(10min.)			
	BE [eV]	FWHM [eV]	area frac. [%]	BE [eV]	FWHM [eV]	area frac.[%]	Δarea frac.
a) O=C	533.55	1.58	44.61	533.69	1.58	52.5	+7.89 %
b) O-C	535.15	1.6	45.34	535.07	1.6	39.66	-5.68 %
% C			71.56			70.8	
% O			28.44			29.2	
O/C			0.398			0.412	

Table 5.5: Fit parameters for the deconvoluted O1s emission spectra, as well as the carbon and oxygen concentration for PMMA after and prior to the UV irradiation for 10min. in ambient atmosphere.

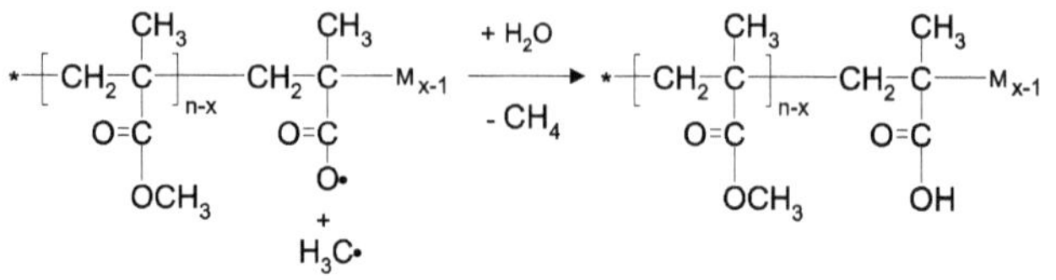

Figure 5.18: Reaction mechanism in the PMMA near surface layer leading to the formation of hydroxyl groups during UV irradiation, as suggested by Wei et al. [61].

formation of hydroxyl groups at the dielectric interface could not be demonstrated by the conducted XPS measurements, owing to the limited resolution of the equipment. The difference in binding energy between O-C and O-H could not be resolved, since the difference in electro negativity between carbon (ΔEN = 2.5) and hydrogen (ΔEN = 2.2) is only 0.3 [59, 89]. To account for the formation of hydroxyl groups in the PMMA near surface layer during the UV modification, a publication by Wei et al. [61] is considered. Here, a similar UV modification of selected PMMA substrates was conducted for a time frame of 30 minutes in ambient atmosphere, prior to an XPS analysis. As for the experiments described above, a radiation wavelength of 254nm with an optical power of $15 \frac{mW}{cm^2}$ was utilized. In order to obtain an indication for the formation of hydroxyl groups by PES, Wei et al. exposed their samples to Tl(OEt), which selectively reacted with available -OH groups to form -OTl. Due to the difference in electro negativity between thallium (ΔEN = 1.4) and carbon (ΔEN = 2.5) of 1.1, the formation of -OTl groups in the O1s emission spectrum could be detected. Indeed, by comparing UV modified and pristine PMMA samples, a significant relative increase in -OTl could be substantiated for the UV exposed substrates, indicating the formation of hydroxyl groups in a near surface layer of the polymer film. The proposed reaction mechanism for the formation of -OH is schematically illustrated in Figure 5.18, suggesting the increase to be the result of a carboxylic acid formation in the PMMA side chain. Here methyl and carboxyl radicals are formed during the UV irradiation, leading to a reaction with atmospheric water to yield methane and the polymer bound carboxylic acid. In accordance with the work at hand, Wei et al. observed, in addition to the relative increase in -OH, the same relative increase in the O=C oxygen component with respect to the O-C component. This is suggested to be the result of an ester side chain scission in the PMMA repeating chain. Since both reaction mechanisms are expected to occur simultaneously, it is suggested, that the conducted UV exposure of PMMA leads to a formation of hydroxyl groups in the polymer near surface layer.

Influence on pentacene topography

In the following, the influence of UV irradiated PMMA insulators on the pentacene topography is investigated, using non contact atomic force microscopy. For this comparative study, a 50nm pentacene layer is deposited onto pristine SiO_2 and PMMA layers as well as onto PMMA thin films, exposed to UV radiation for respective times of 5 and 10 minutes. The deposition onto room temperature

5.2 UV modified Polymethylmetacrylate

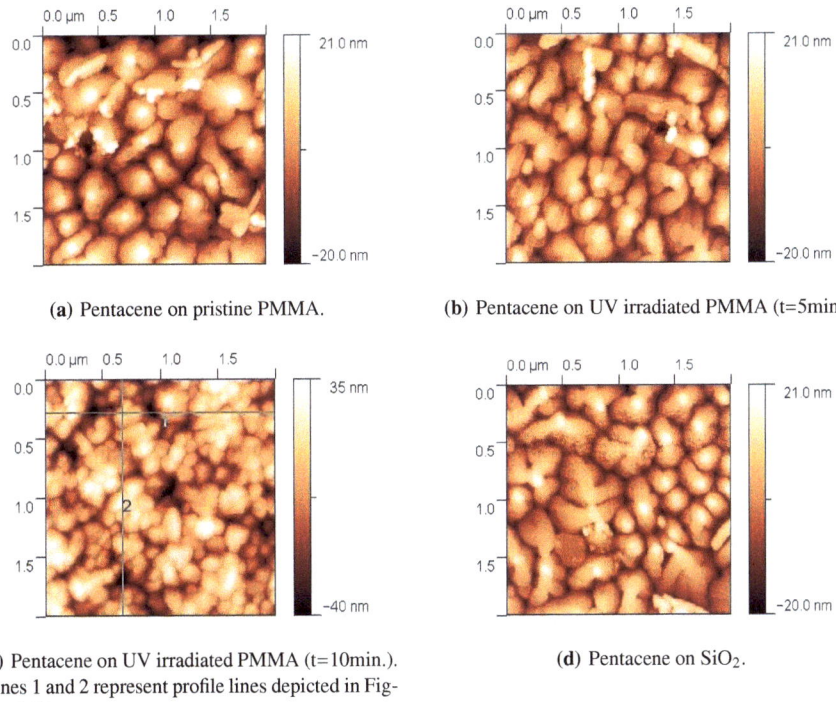

(a) Pentacene on pristine PMMA.

(b) Pentacene on UV irradiated PMMA (t=5min.).

(c) Pentacene on UV irradiated PMMA (t=10min.). Lines 1 and 2 represent profile lines depicted in Figure 5.20.

(d) Pentacene on SiO$_2$.

Figure 5.19: AFM micrographs of pentacene deposited onto different dielectric layers, obtained in the non contact mode.

substrates was conducted, using physical vapor deposition (PVD) at a rate of $2\frac{\text{Å}}{\text{s}}$ as well as a chamber base pressure below 10^{-6}mbar.

The 2 x 2μm^2 micrographs of these measurements are illustrated in Figures 5.19(a)-5.19(d), with the extracted data being summarized in Table 5.6. Considering the topography of the samples, it is found, that the pentacene crystallite form changes from a dendritical to a spherical structure with increasing PMMA UV exposure time. This coincides with a decrease in pentacene grain size[17] average from $\approx$ 499nm for pristine PMMA down to an average of $\approx$ 143nm for PMMA exposed to 10 minutes of UV radiation. The observation is accompanied with an increase in rms roughness from R_q = 8.38nm up to a value of R_q = 14.93nm. In part responsible for this increase in surface roughness is the deviation of the pentacene layer profile with respect to its median of at least -20.79nm for all of the investigated samples, as summarized in Table 5.6. Pentacene thin films deposited onto 10 minute UV irradiated PMMA samples even deviate by almost 50nm, which is in the order of the layer thickness. This is emphasized in Figure 5.20, illustrating line scans of the pentacene topography, as

---

[17] Here the largest axis of the respective pentacene grains is considered as the grain size.

derived using Figure 5.19(c) along the indicated profile lines one and two. By taking the standard OFET top contact device structure into account, it can be concluded, that no planar source-drain metalization / pentacene interface is given.

In general, when only considering pentacene-substrate interaction, the pentacene growth depends on interface roughness [66, 90] and has also been indicated to be influenced by the substrate polarity [66, 91]. However, due to an obtained PMMA surface roughness of $R_q < 1$nm for pristine and UV modified thin films, the observed changes in the pentacene topography cannot be ascribed to this parameter as substantiated by Shin et al. [90]. By taking the decrease in water contact angle with increasing UV exposure time into account, it seems possible to correlate the obtained pentacene topography changes with an increase in substrate polarity. However, by considering the pentacene topography obtained on top of a SiO_2 dielectric, as depicted in Figure 5.19(d), this approach appears to be invalid. This is concluded, since the pentacene topography on top of SiO_2 is similar to the topography obtained on top of pristine PMMA samples, while the surface roughness and contact values for pristine SiO_2 and 10 minute UV exposed PMMA substrates yield the same values (rms < 0.4nm, $CA_{SiO_2} \approx 36°$). Possibly, the changes in the pentacene topography can be ascribed to the formation of low molecular weight ashes on the PMMA surface during the photooxidation process [61, 92].

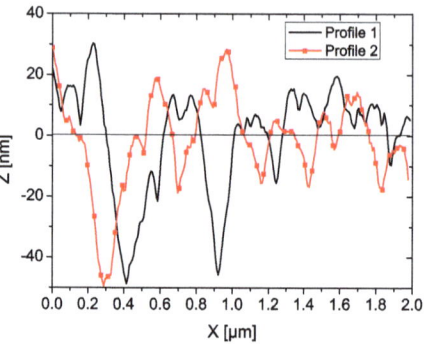

Figure 5.20: Pentacene profile of the profile lines indicated in Figure 5.19(c)

In the current subsection, the influence of UV irradiation in ambient atmosphere on PMMA interface properties was investigated. It could be substantiated, that the interface roughness of PMMA is not influenced by the UV exposure, while the water contact angle is significantly degraded from 80°, down to a value of 30°, during a UV exposure time of 15 minutes. The degradation in water contact angle is suggested to be the result of an increase in oxygen containing polar groups in the PMMA near surface layer, due to photooxidation. These polar groups could be identified as keto

	PMMA	PMMA	PMMA	SiO_2
Substrate characteristica				
UV exposure [min.]	0	5	10	-
Contact angle [°]	81	41	35	36
Rms roughness [nm]	0.3	0.3	0.3	0.4
Pentacene morphology				
Grain form	Dendritical	Dendritical	Spherical	Dendritical
Grain size [nm]	499	407	143	488
Rms roughness [nm]	8.38	7.27	14.93	7.33
Median deviation [nm]	-26.66	-20.79	-49.92	-22.15

Table 5.6: Summary of the data extracted from the interface analysis of SiO_2, PMMA and Pentacene, as discussed in section 5.2.1.

5.2 UV modified Polymethylmetacrylate

and hydroxyl groups, using the result of PES measurements in accordance with literature [61]. Furthermore, it could be demonstrated, that the topography of pentacene thin films deposited onto UV modified PMMA layers is significantly influenced by the exposure time of the PMMA layer. With increasing PMMA irradiation time, the pentacene topography is changed from a large grained dendritical structure for pristine PMMA layers to small grained spheres. The topography of pentacene layers deposited onto UV modified thin films exhibit large cavities, with depths corresponding to almost the entire layer thickness.

5.2.2 Influence of UV modified PMMA gate dielectrics on OFET transport properties

In the preceding subsections, it could be reasoned, that the UV irradiation of PMMA leads to the formation of keto as well as hydroxyl groups in a near surface layer of the dielectric without changes in the interface roughness (R_q < 1nm). In accordance with literature [29, 43], such functional groups have been demonstrated to represent electron traps for pentacene charge carrier transport. In the following subsection, the influence of a UV modified PMMA dielectric interface on the OFET charge carrier transport is explored. For this investigation, $\approx$ 119nm thick films of PMMA are deposited onto the standard p^{++} – Si / SiO_2 substrates by spin coating, using a 2%wt THF solution. Subsequently, the PMMA layer is UV irradiated for 10 minutes in ambient atmosphere. A 50nm pentacene layer is then deposited by PVD. The OFET device structure is finalized by the PVD of Ca source-drain contacts with a thickness of 100nm. The exact process parameters are listed in Appendix D.

Electrical OFET Performance

In the following, the transistor characterization is conducted by driving measurement cycles in the hole as well as the electron accumulation mode. A measurement cycle represents the variation of the drain-source and the gate-source voltages between 0V and 80V, with a respective stepwise increase of $|\Delta V_D|$ = 1V and $|\Delta V_G|$ = 20V.

Illustrated in Figure 5.21(a) is the output characteristic of a UV modified transistor in the electron and hole accumulation mode. By taking the open dot scatter plot in the hole accumulation into account, which represents a single measurement at V_G = -80V, only a negligible hole current is obtained. This is expected, due to an insufficient matching of the Ca metal work function and the HOMO level of the organic semiconductor. The closed dot scatter plot, illustrated in the electron accumulation, represents the current voltage characteristic at V_G = 0V during the first transistor measurement cycle. Due to a gate voltage of V_G = 0V, no electron current is obtained for this characteristic. However, with an increase in gate voltage, the drain current remains zero. The impeded electron transport is suggested to be the result of keto and hydroxyl groups, introduced to the dielectric interface by UV radiation. It needs to be pointed out, that at V_G = 0V, a low increase in drain current is registered for V_D $\approx$ 80V. This current can be significantly increased, as indicated by the closed triangle scatter plot of Figure 5.21(a), if the OFET is exposed to an electrical cyclic conditioning in the electron accumulation. The obtained increase in drain current could be the result of leakage, an unsaturated electron

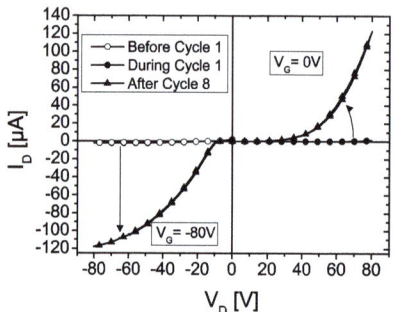
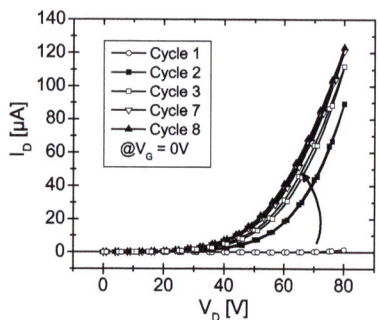

(a) OFET current voltage characteristic for $V_G=0V$ in the electron accumulation and $V_G=-80V$ in the hole accumulation mode.

(b) Output current voltage characteristic in the electron accumulation during the the electrical cyclic conditioning step at $V_G=0V$.

Figure 5.21: Output characteristic for a pentacene OFET comprising a UV modified PMMA dielectric in its pristine and electrical cyclic conditioned state.

current, or be due to a unipolar hole current in the ambipolar range of the electron accumulation, as discussed in section 2.2.1. By taking into account the neglectable gate leakage current, recorded parallel to the measurement (not shown), this type of interpretation of the drain current can be excluded. The unipolar p-type behavior of the electrically conditioned transistor, however, can be confirmed by considering the output characteristic in the hole accumulation. This measurement is depicted by the closed triangle scatter plot at $V_G = -80V$ in the 3rd quadrant of Figure 5.21(a). Indeed, a unipolar p-type drain current is obtained for the measurement, exhibiting hardly any hysteresis. The s-shaped behavior in the linear range of the characteristic is an indication for the expected large injection barrier for positive charge carriers from Ca contacts into pentacene. It needs to be stressed, that the hole injection occurs despite the discussed large misalignment in the Ca metal work function and the pentacene HOMO level of $\approx 2.11eV$.

The electrical cyclic conditioning[18], in order to enable the p-type transistor behavior, is applied by the repeated execution of the measurement cycle in the electron accumulation between 0V and 80V for the respective drain-source and gate-source voltages. Depicted in Figure 5.21(b) is the drain current at $V_G = 0V$ for the individual conditioning steps. Between the first and the second conditioning cycles, a significant increase in the drain current is obtained, which is continuously enhanced after each cycle, until saturation is achieved between the 7th

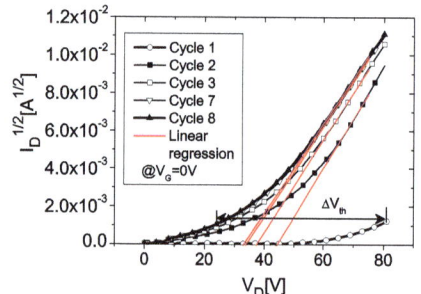

Figure 5.22: Threshold voltage shift between the 1st and 8th electrical cyclic conditioning step for an OFET incoperating a UV modified PMMA gate dielectric.

---

[18] See section 3.2.3 for a detailed explanation

5.2 UV modified Polymethylmetacrylate

or 8th conditioning step. By considering Figure 5.22, the origin of the significant increase in unipolar p-type drain current in the electron accumulation can be understood. In this graph, the square root of the experimental data is compared to the square root of the ambipolar drain current equation, for $\mu_n = 0\frac{cm^2}{Vs}$, $V_G = 0V$, and $|V_D| \geq |V_G - V_{th,h}|$ as shown below:

$$\sqrt{I_D} = \sqrt{\frac{wC_{tot}\mu_h}{2l}}(V_D + V_{th,h} - \Delta V_{th}) \qquad (5.1)$$

Here, $C_{tot} = 10.4\frac{nF}{cm^2}$ represents the resulting total areal capacitance of the SiO_2 / PMMA bilayer dielectric. Apparently, the improvement in hole current can be related to a positive threshold voltage sift of $\Delta V_{th} \approx 60V$ during the first eight conditioning cycles. The observed threshold voltage shift has been derived, using the respective output characteristics in the hole accumulation, and occurs with an almost constant hole mobility. This is deducted from the constant inclination of the linear regression fits during the shift in threshold (Figure 5.22). Here, a trapping of electrons (n_t) within the UV modified PMMA near surface layer is suggested to be the cause for the observed ΔV_{th}. During the cyclic conditioning in the electron accumulation, negative charge carriers are accumulated at the dielectric interface and localized in electron traps, generated by the UV irradiation. This leads to a change in the electric field distribution in the transistor channel and therefore to a large threshold voltage shift for both electrons and holes. In a crude first order approximation, the trapped charge carrier density n_t is estimated to be:

$$n_t = \frac{\Delta V_{th}C_{tot}}{q} = 3.9 * 10^{12}\frac{1}{cm^2} \qquad (5.2)$$

The value of $3.9 * 10^{12}\frac{1}{cm^2}$ represents the lowest possible estimate and is only valid, if the negative charge is trapped directly at the dielectric interface. For a trapping in the volume of the PMMA, the trapped charge carrier density must be higher. By considering Figures 5.21(a) and 5.21(b), the negligible hysteresis of the current voltage characteristics suggests the absence of recombination between trapped negative charge carriers and mobile positive charge carriers during the transistor operation. This holds true, even if the OFET is operated in the hole accumulation mode. The absence of recombination indicates, that the localized electrons are trapped in a near surface layer of the PMMA insulator, which isolates the trapped negative charge from the mobile positive charge in the transistor channel.

Charge carrier injection

It has been demonstrated, that the polarity of a unipolar n-type OFET can be switched to unipolar p-type by negatively charging its UV modified PMMA dielectric. The injection of holes occurred despite a pronounced injection barrier, which is due to the energetic difference in metal work function ($E_A = 2.9eV$) and the HOMO level ($E_{HOMO} = 5.01eV$) of the intrinsic organic semiconductor. In the following interface dipoles, classical Fowler-Nordheim (FN) tunneling, as well as Schottky barrier lowering are considered for a discussion of this injection phenomenon.

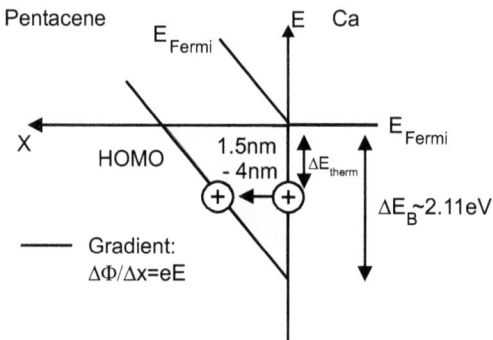 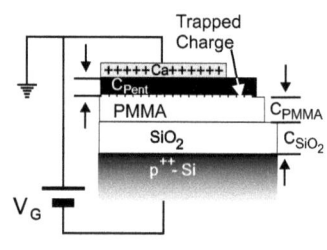

(a) Hole tunnel injection from a Ca electrode into pentacene. The injection is approximated using a triangular barrier at electric field strengths of $1.94 * 10^8 \frac{V}{m}$ and $3.16 * 10^8 \frac{V}{m}$.

(b) Transistor cross section below the source or drain electrode, incorporating a charged PMMA dielectric.

Figure 5.23: Field enhanced tunnel injection.

- The formation of **interface dipoles** at metal / organic semiconductor contacts has been subject to a lot of recent interest in order to explain charge carrier injection despite a pronounced mismatch between the metal work function and the HOMO / LUMO transport levels [93–95]. The formation of an interface dipole results in a shift of the vacuum level and therefore a change in the energetic position of the HOMO / LUMO levels with respect to the Fermi energy. In dependence of the norm or the direction of the shift to lower or higher binding energies, such a dipole is beneficial for the injection of either electrons or holes, due to a reduction in injection barrier height. For the case at hand, however, where pentacene / Ca interfaces are considered, Watkins et al. [95] have published a vacuum level shift of $\Delta E_{vac} \approx 0.35$eV, that enhances the required thermal activation for the injection of positive charge carriers instead of lowering it. It is therefore suggested, that the observed injection of holes is not the result of a dipole shift at this interface.

- The negative charging of a UV modified dielectric results in a change in the electric field distribution in the transistor channel. This may lead to a **tunnel injection** of holes, due to an enhancement in the electric field strength, as is evaluated in the following by taking into account the FN theory. For these considerations, a triangular injection barrier is assumed, as indicated in Figure 5.23(a). This approach seems valid, since the resulting electric field between the trapped negative charges and the transistor source or drain electrodes should, in part, force the accumulated space charge at the Ca / semiconductor contact out of the active layer and lead to a tilt in the semiconductor transport levels [19]. Furthermore, it is assumed, that the injection of positive charges has not yet occurred and the thermal generation of charges in the transistor channel can be neglected. The electric field strength E in the transistor channel is approximated by

---

[19] The linear potential ϕ gradient of the transport levels in dependence of the electric field strength E and distance x is given by $\frac{\Delta \phi}{\Delta x} = eE$.

5.2 UV modified Polymethylmetacrylate

assuming the trapped negative charge to be localized directly at the dielectric / semiconductor interface:

$$E = \frac{V_G}{\frac{\epsilon_{r,Pent}}{\epsilon_{r,SiO_2}}d_{SiO_2} + \frac{\epsilon_{r,Pent}}{\epsilon_{r,PMMA}}d_{PMMA} + d_{Pent}} + \frac{n_t e}{d_{Pent}(C_{Pent} + \frac{C_{SiO_2}C_{PMMA}}{C_{SiO_2}+C_{PMMA}})} \quad (5.3)$$

The first part of equation 5.3 describes the contribution to the electric field strength of the applied gate voltage, while the contribution of the trapped charge n_t is given by the second part of the equation. Here, d_{SiO_2}, d_{PMMA} and d_{Pent} represent the respective SiO$_2$, PMMA and pentacene layer thickness. The dielectric constant of the respective layers is described by $\epsilon_{r,PMMA} = 3.5$ for PMMA, $\epsilon_{r,Pent} = 3.7$ for pentacene [54] and $\epsilon_{r,SiO_2} = 3.9$ for SiO$_2$. The parameter C_{Pent} describes the pentacene capacity between the trapped charge and the source electrode, while C_{PMMA} and C_{SiO_2} describe the capacitance of the SiO$_2$ and PMMA insulator layers, as is indicated in Figure 5.23(b).

It has been demonstrated above[20], that source or drain metalization spikes extend through almost the entire pentacene layer thickness. Consequently, by assuming a homogeneous pentacene layer thickness of either $d_{Pent} = 1.6$nm or $d_{Pent} = 50$nm, the highest and lowest estimate in electric field strength below the source-drain electrodes can be calculated.

By using equation 5.3, pentacene layer thicknesses of 50nm and 1.6nm as well as gate voltages of 0V and -80V, the thermal activation energies required to tunnel either 1.5nm or 4nm[21] distances through the triangular injection barrier can be calculated. This is indicated in Figure 5.23(a) for the case of an elevated electric field strength. The resulting energies E_{therm} are summarized in Table 5.7. For applied gate voltages of -80V as well as a tunnel distance of 4nm, reasonably small thermal activation energies of 0.81eV ($d_{Pent} = 50$nm) and 0.36eV ($d_{Pent} = 1.6$nm)for tunnel injection are found. For all other parameters, however, the thermal activation required for tunneling is well above 1eV, which seems too high in order to enable tunnel injection. It is therefore concluded, that the experimentally obtained transistor currents, in the order of 10^{-5}A, are in part due to a tunneling current. However, it seems unlikely, that the entire transistor current results from a tunneling effect.

Pent. thickness [nm]	50		1.6	
$\|V_G\|$ [V]	0	80	0	80
$E_{therm,1.5nm}$ [eV]	1.95	1.62	1.83	1.45
$E_{therm,4nm}$ [eV]	1.68	0.81	1.37	0.36

Table 5.7: Thermal activation required for tunnel injection through 1.5nm and 4nm distances at different field strengths an pentacene layer thicknesses.

---

[20] See Figure 5.20.
[21] The tunnel distance of 4nm is suggested as an upper limit estimate for tunnel injection.

- Due to the high electric field strength available in the transistor channel, a **Schottky barrier lowering**, as described by equation 5.4 [96], may allow for the thermionic injection of holes at the Ca / pentacene interface.

$$\Delta E_B = q \sqrt{\frac{eE}{4\pi\epsilon_0\epsilon_{r,pent.}}} \tag{5.4}$$

The following estimate is conducted under the assumption of constant electric field strengths of either $4.4 * 10^8 \frac{V}{m}$ or $3.25 * 10^8 \frac{V}{m}$. These values are obtained using equation 5.3 for respective pentacene layer thicknesses of 1.6nm and 50nm at an applied gate voltage of -80V. A reduction in the hole injection barrier height of $\Delta E_B \approx 0.41\text{eV}$ or $\Delta E_B \approx 0.36\text{eV}$ is calculated. For the case of a homogeneous 1.6nm thick pentacene layer, this would result in an hole injection barrier of $E_B \approx 1.7\text{eV}$. Considering, that the resulting injection barrier remains high, it seems unlikely, that the discussed p-type current voltage characteristics are significantly influenced by a Schottky barrier lowering.

Taking into account the above estimates, it is clear, that the observed hole injection from Ca contacts into pentacene can only be rudimentarily explained. While it seems safe to conclude, that the observed hole current is not due to only one of the above discussed aspects, it is unsure, whether all three aspects combined would lead to the experimentally demonstrated magnitude in transistor drain current. This seems to be true, since the discussed interface dipole at the Ca / pentacene contact is suggested to have no beneficial effect on the hole injection. Interface effects, that have not been considered, could possibly lead to an additional improvement in the charge carrier injection, such as variations in the metal work function or a further field enhancement at metalization spikes.

Degradation in transistor performance

The p-type characteristic of UV modified pentacene OFETs exhibits only a negligible current hysteresis during the measurement. This indicates an insignificant recombination between trapped electrons and mobile positive charge carriers during the recording of a device characteristic. However, the transistor performance is subject to degradation over time, as indicated in Figure 5.24. Here, the degradation in maximum drain current $I_{D,max}$ recorded for $V_G = -80\text{V}$ is illustrated by the closed squared scatter plot, as obtained, using output characteristic measurements. The dotted vertical lines of Figure 5.24 indicate performed transfer characteristic measurements. An almost linear degradation is obtained during the time frame of the experiment ($t = 50$ min.), which is independent of the number of successive measurements. This is conjectured, using

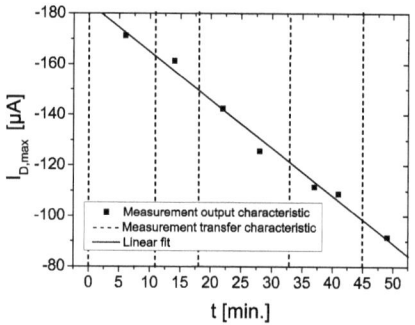

Figure 5.24: Degradation in $I_{D,max}$ for a UV-modified p-type OFET over time. The squares represent values derived from output characteristic measurements. The vertical dotted lines indicate the measurement of a transfer characteristic. All measurements were conducted in the hole accumulation.

5.2 UV modified Polymethylmetacrylate

t [min.]	$V_{th,p}$ [V]	$\mu_h \frac{cm^2}{Vs}$
0	-49.42	0.43
11	-52.50	0.44
18	-55.20	0.50
33	-57.76	0.49
45	-60.10	0.49

Table 5.8: Degradation in OFET threshold and mobility over time.

the experimental fact, that the obtained degradation in $I_{D,max}$ varies insignificantly from measurement sweep to measurement sweep, while exhibiting an almost linear dependence with regard to time. Illustrated in Table 5.8 are the parameters $V_{th,p}$ and μ_h, as derived using the indicated transfer characteristic measurements. As expected, when considering the degradation in transistor drain current, the norm in threshold voltage is increased over time. The extracted hole mobility, however, remains almost constant. This result is in line with the observed improvement in OFET p-type performance, due to the cyclic electrical conditioning, as discussed above. By taking the negative threshold voltage shift of about 10.7V during the time dependent measurement into account as well as equation 5.2, a discharge of $\Delta n_t \approx 0.69 * 10^{-12} \frac{1}{cm^2}$ charges through the semiconductor and the respective source-drain electrodes is estimated for a time frame of 45 minutes.

To further investigate the OFET discharge process without possible negative influences of the glove box environement, such as solvent vapors, rest oxygen or water, the following experiment is conducted. A UV modified transistor is introduced into a UHV chamber with a base pressure of $< 10^{-10}$ mbar, which allows for an electrical contact of the investigated device. The transistor cross section is identical with the standard OFETs, however, the device differs with respect to its channel dimensions. Here, a $\frac{w}{l}$ ratio of 10 is chosen. After its cyclic conditioning in the UHV, the OFET is characterized immediately, then after four weeks in its electrically degraded and finally in its electrically reconditioned state. Depicted in Figure 5.25 are the resulting output characteristic measurements at an applied gate voltage of $V_G = -80V$. It can be concluded, that a linear dependency of the discharge process, as implied by the degradation in transistor drain current illustrated in Figure 5.24 is not a valid approximation. A linear dependency of the discharge process[22] would result into a complete discharge of the localized electrons within a time frame of approximately 4 hours. However, within a time frame of 4 weeks (11th Dec. 06), the drain current $I_{D,max}$ is only degraded by a factor of ≈ 5.6. and is not completely inhibited, as one would expect for a completely discharged dielectric.

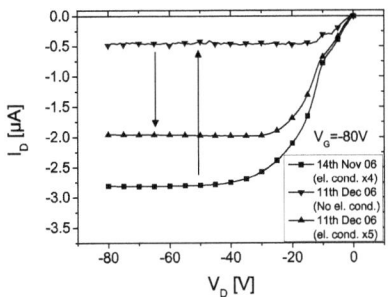

Figure 5.25: Degradation as well as reconditioning of a UV modified OFET in the UHV ($< 1 * 10^{-10}$ mbar), depicted for the case of output characteristics recorded at $V_G = 80V$.

---

[22] $\frac{\Delta n_t}{dt} \approx -0.15 * 10^{11} \frac{charges}{minute cm^2}$

The observed transistor degradation is proposed to be due to a thermal discharge of the UV modified and charged PMMA dielectric. Thermal discharge is generally considered to exhibit an exponential [56] and not a linear behavior. Furthermore, by considering Figure 5.25, it is demonstrated, that the p-type performance of the electrical degraded transistor can be reenhanced by exposing the device to an additional electrical cyclic conditioning. Under the assumption, that most electrons are localized in hydroxyl groups, this allows the following two conclusions about the trapping mechanism of electrons in hydroxyl groups:

- The electron trapping mechanism in -OH groups is suggested by Chua et al. [29] to be due to the dissociation of a hydrogen atom. Therefore, once an electron is discharged from such a trap, it is inactive, unless the proton is reattached to the remaining oxygen radical. The observed possibility, to recharge the PMMA dielectric, as indicated in Figure 5.25, may lead to the suggestion, that not all of the available electron traps are filled during the cyclic electrical conditioning. Therefore, one could ascribe the observed reenhancement in p-type performance to a filling of surplus trap states. This seems possible, especially when considering the density of hydroxyl groups available on UV modified PMMA interfaces in the order of $4 * 10^{14} \frac{1}{cm^2}$, as discussed by Wei et al. [61]. This value is two orders of magnitude larger than the value estimated for the trapped charge carrier density of $3.9 * 10^{12} \frac{1}{cm^2}$.

- However, under the assumption, that all of the available electron traps at the dielectric interface are filled during the cyclic electrical conditioning, the experimental fact that, the OFET p-type characteristic can be reenhanced by an additional conditioning step, implies, that the trapping mechanism, as suggested by Chua et al. [29], is invalid. It is unlikely, that the hydrogen atom, once dissociated during the electron trapping mechanism, remain in the vicinity of the trapped charge and is reattached to the oxygen radical upon its discharge.

In the preceding section, it was demonstrated, that functional end groups in the form of keto and hydroxyl groups can be introduced to the near surface layer of a PMMA dielectric by exposing the insulator to UV radiation in ambient atmosphere. It was shown, that this type of dielectric interface treatment allows for a change in the polarity of an otherwise unipolar n-type pentacene OFET to unipolar p-type by exposing the transistor to a cyclic electrical conditioning in the electron accumulation. The change in polarity is due to large threshold voltage shift of $\Delta V_{th} \approx 60V$ for electrons and holes, which is proposed to be the result of negative charges trapped by keto and hydroxyl groups at the dielectric interface. The trapped charge leads to change in the electric field distribution in the transistor channel, resulting in a field enhanced injection of holes. This occurs despite a suggested energy barrier of 2.11eV. At the same time, the electron current in the transistor channel is suppressed, due to coulombic repulsion. The p-type characteristic of a UV modified OFET is subject to degradation over time, it is, however, independent of the presence of holes. This suggests, that the trapped negative charges are isolated from mobile positive charges present in the transistor channel during measurements in the hole accumulation.

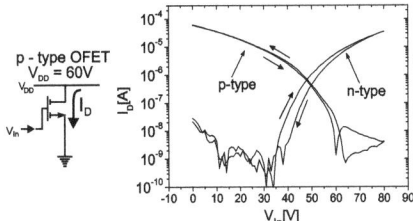 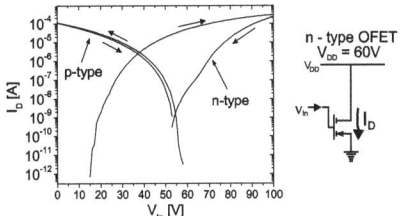

(a) Characteristics of complementary OFETs, realized on a PMMA dielectric. The PMMA dielectric of the p-type transistor is UV modified, otherwise the transistor is identical to the n-type device.

(b) Transfer characteristics for p- and n-type pentacene OFETs, realized on SiO_2. The insulator area for the n-type transistor is Ca passivated.

Figure 5.26: Transfer characteristic of complementary OFETs. In analogy to an inverter structure, the source contact of the p-type OFET is held at a potential of 60V for both plots.

5.3 Applications

In the previous sections, two dielectric interface modifications were demonstrated by which unipolar p- and n-type transport can be achieved for pentacene OFETs. In the following, these approaches are applied to the fabrication of organic complementary metal oxide semiconductor (O-CMOS) devices, namely inverter stages. A CMOS inverter stage consists of two FETs of complementary polarity (p- and n-type), sharing a common gate as well as a common drain electrode, to form the respective inverter in- and output electrodes. The equivalent circuit of the realized inverter structures is shown in the inset of Figures 5.27(a) and 5.27(b).

The process parameters for the realization of the complementary OFET types as well as a description of their cross section have been summarized in sections 5.1 and 5.2. The exact process parameters are discussed in detail in Appendix D. In order to characterize the respective transistors, prior to their use in an inverter structure, the OFETs are measured in accordance with the equivalent circuits as depicted next to the Figures 5.26(a) and 5.26(b). Therefore, V_G is addressed as V_{In} in the following, and the source potential of the p-type OFET is held constant at $V_{DD} = 60$V. The parameter V_{DD} represents the supply voltage of the inverter measurement.

Using the UV modification technique for the realization of an organic CMOS inverter stage, as discussed in section 5.2.2 [44, 97], the SiO_2 / PMMA dual layer dielectric of the designated p-type OFET is irradiated by UV radiation in ambient atmosphere. By exposing the completed transistor to a cyclic electrical conditioning, the device polarity is switched from unipolar n-type to p-type. The respective transfer characteristics of the complementary OFETs in the hole and electron accumulation are depicted in Figure 5.27(a). Both devices exhibit only a very low current hysteresis. Electron and hole mobilities were determined to be $\mu_e = 0.78 \frac{cm^2}{Vs}$ and $\mu_h = 0.11 \frac{cm^2}{Vs}$ respectively; an $\frac{On}{Off}$ ratio exceeding 10^4 was obtained for both devices. The respective threshold voltages are $V_{th,e} = 46$V and $V_{th,p} = -27.5$V. The obtained balanced transistor parameters allowed for the realization of an O-CMOS inverter stage. The inverter characteristic is illustrated in Figure 5.27(a) with the equivalent circuit used for the characterization being depicted by its inset. The inverter exhibits stable operation below

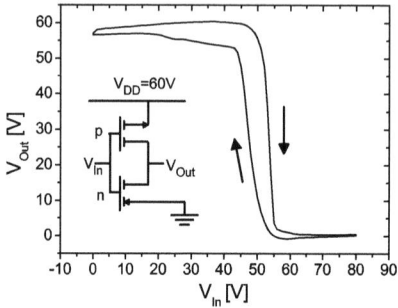
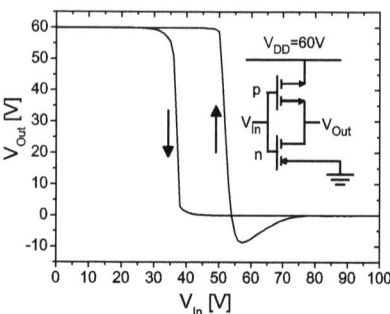

(a) This O-CMOS inverter has been realized using the PMMA dielectric UV modification approach. It exhibits a max. gain of 17.

(b) This O-CMOS inverter has been realized using the Ca passivation technique. It exhibits a high gain of 24.

Figure 5.27: Organic CMOS inverter stage transfer characteristic.

its supply voltage of V_{DD} = 60V as well as a maximum gain of 17. The observed hysteresis of the inverter is ascribed to the current hysteresis of the utilized p- and n-type pentacene OFETs.

As discussed in section 5.1, the Ca passivation technique can also be used for the realization of an organic CMOS inverter stage, as demonstrated by Ahles et al. [83, 98]. In the following, the experimental data of Ahles et al.[23] is presented as a comparison to the inverter experiment described above. Here, a standard p^{++} – Si / SiO$_2$ substrate is selectively covered with a 6Å layer of Ca. The n-type OFET is realized on the Ca passivated area by the use of Ca source-drain contacts. The p-type OFET is realized on the same substrate in the area where pristine silicon dioxide is available, using an Au source-drain metalization. The transfer characteristics of the individual p- and n-type OFETs are illustrated in Figure 5.26(b). The n-type OFET has been electrically cyclic conditioned prior to the measurement in order to enhance the n-type performance of the OFET. Using these characteristics, an electron and hole mobility of $\mu_e = 0.11 \frac{cm^2}{Vs}$ and $\mu_h = 0.10 \frac{cm^2}{Vs}$ as well as $\frac{On}{Off}$ ratios > 10^5 were determined. The respective threshold voltages are $V_{th,e}$ = 34V and $V_{th,p}$ = -20V. While the p-type OFET exhibits only a negligible current hysteresis, a large hysteresis is obtained for the n-type OFET between the forward and reverse voltage sweep of V_{In}. This is ascribed to residual electron traps at the dielectric / semiconductor interface for the Ca passivated n-type OFET. Owing to balanced charge carrier transport properties of the respective transistors, an O-CMOS inverter stage was realized by connecting the devices as illustrated by the inset of Figure 5.27(b). The inverter exhibits a stable operation below its supply voltage of V_{DD} = 60V as well as a maximum gain of 24. The obtained hysteresis in the inverter characteristic reflects the current hysteresis of the n-type OFET depicted in Figure 5.26(b).

While for both the UV modification and Ca passivation technique O-CMOS inverter stages could be realized, the UV modification technique seems better suited for technological application. This is

---

[23] The presented data is used with the kind approval of Ahles et. al.

5.3 Applications

suggested to be due to the much lower current hysteresis of the n-type OFET realized on a PMMA dielectric as compared with the current hysteresis of the n-type transistor realized on a Ca passivated SiO_2 insulator. Furthermore, the circumstance that unipolar n- and p-type OFETs can be realized with an identical device structure by the use of the UV modification technique, makes the approach potentially attractive for product applications.

Chapter 6

OFET threshold tuning by the use of an electret

In section 5.2, a technique was demonstrated by which the polarity of a unipolar n-type pentacene OFET can be switched to unipolar p-type without modifying its device cross section. The polarity inversion is due to a large positive threshold voltage shift for both electrons and holes. The threshold voltage shift is obtained by negatively charging the gate dielectric, which leads to a suppression of the electron current by coulombic repulsion and allows for a hole current by field enhanced charge carrier injection from Ca electrodes despite a pronounced injection barrier. In the current chapter, the complementary experiment is investigated by using the electret characteristic of a polymethylmetacrylat (PMMA) dielectric. The aim of this experiment was to obtain a negative threshold voltage shift in the transistor current voltage characteristic by positively charging the gate insulator and therefore to influence the polarity of an otherwise p-type pentacene OFET. Using this technique, it was attempted to substantiate the field enhanced injection of electrons out of Au contacts into pentacene, instead of the previously discussed hole injection out of Ca electrodes. Technologically, such an approach is more interesting, since Au electrodes are stable in ambient atmosphere.

In general, the charging of an electret such as PMMA, if electrically contacted by electrodes on both sides, is the result of dipole orientation, space charge separation or, at high electric field strengths, the result of injection and storage of excess charge carriers [30, 99, 100]. While the charge carrier injection is mainly dependent on the absolute value of the electric field strength, elevated substrate temperatures enhance the conductivity of the electret, thereby allowing the charging of energetically deep traps in the bulk of the dielectric. The electrostatic field, resulting from an electret formed in such a fashion, is therefore the outcome of a superposition of the electrostatic fields, resulting from the injected homocharge and separated or oriented heterocharge.

The experiments are conducted, using the standard pentacene OFET device structure, as discussed in section 3.2.2. The OFETs comprise a SiO_2 / PMMA dual layer dielectric with respective thicknesses of 200nm and 119nm as well as a 100nm Au source-drain metalization. In order to positively charge the polymer insulator by injected excess charges, the completed OFET device structure is exposed to several forming steps at high electric field strengths of E_{Form}= -1.34$\frac{MV}{cm}$, -2.01$\frac{MV}{cm}$ and -2.68$\frac{MV}{cm}$. The device forming is conducted at the glass temperature of PMMA (T_{Form} = 108°C [56]). The experimental setup is discussed in detail in section 3.2.3.

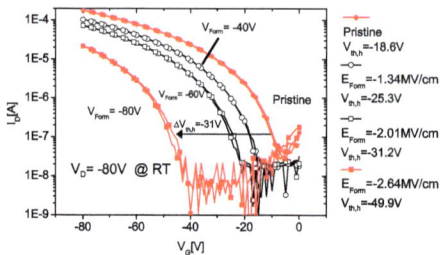 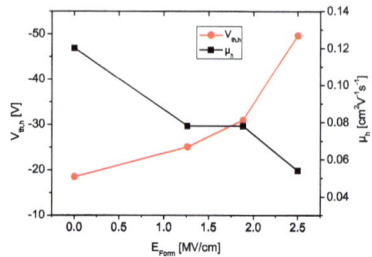

(a) Transfer characteristics for $V_D = -80V$ at room temperature after the individual forming steps.

(b) Threshold voltage and mobility development in dependence of the applied forming steps.

Figure 6.1: Influence of the respective forming steps on the OFET device characteristic.

The influence of the respective forming steps on the OFET current-voltage characteristic is demonstrated in Figures 6.1(a) and 6.1(b). The transfer characteristics of an investigated OFET in its pristine state and after the subsequent forming steps are illustrated in Figure 6.1(a) as a log(I_D) over V_G plot. All of the measurements were conducted at room temperature and are depicted for an applied drain voltage of $V_D = -80V$. The development of the transistor threshold voltage and charge carrier mobility during the respective forming steps is depicted in Figure 6.1(b). The parameters were extracted using the transfer characteristics illustrated in Figure 6.1(a). For the initial forming step, using an electrical field strength of $E_{Form} = -1.34 \frac{MV}{cm}$, a threshold voltage shift of -6.7V from -18.6V to -25.3V is obtained. The additional subsequent forming steps, conducted at $E_{Form} = -2.01 \frac{MV}{cm}$ and $-2.68 \frac{MV}{cm}$, lead to an increase in $V_{th,h}$ to values of -31.2V and -49.9V. This results in a total $\Delta V_{th,h}$ of -31.3V. Associated with the observed threshold voltage shift is a reduction in hole mobility μ_h between the transistor in its pristine and its respective formed states from $\mu_h = 0.12 \frac{cm^2}{Vs}$ to $= 0.054 \frac{cm^2}{Vs}$. The degradation in hole mobility is proposed to be the result of possible changes in the pentacene morphology, due to elevated substrate temperatures during the forming step. The conclusion is based on the result of a temper experiment described in section 5.1.4. In this experiment, a unipolar p-type pentacene OFET, realized on a pristine SiO_2 dielectric, was exposed to an annealing step at 160°C. In analogy to the experiment discussed above, the hole mobility was degraded. This was ascribed to the negative effect of heat induced pentacene morphology changes. However, the pronounced negative shift in $V_{th,h}$, as illustrated in Figure 6.1(b) is not ascribed to the influence of pentacene morphology changes. The shift is proposed to be the result of the electret charging. This conclusion is supported by the experimental fact, that the threshold voltage shift is reversible, as illustrated in Figure 6.2. The open and filled squared scatter plots represent the OFET characteristic in its pristine state and after its second forming step, emphasizing the negative threshold voltage shift. The open dot and open triangle scatter plots represent the transfer characteristic, obtained during the rising temperature slope of the third forming step at $T = 50°C$ and $T = 100°C$. While the threshold voltage remains almost unchanged for the measurement at $T = 50°C$, a significant shift in threshold voltage toward the threshold voltage of the pristine device is obtained for the measurement at $T = 100°C$. The shift is attributed to thermally activated discharge during the measurement of the transfer characteristic. Once the positive charge is thermally activated, it is discharged through the pentacene and the source-drain metalization.

Illustrated in Figure 6.3(a) is the displacement current during the first forming step ($V_{Form} = 40V \Rightarrow E_{Form} = -1.34\frac{MV}{cm}$), as recorded using the gate electrode. A decay of the displacement current down to a value of zero after 15 minutes is found. The inset of Figure 6.3(a) represents the total amount of ac-

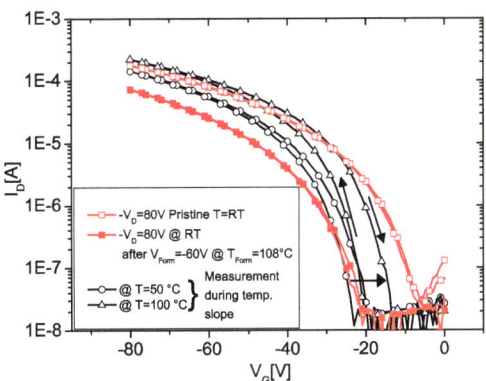

Figure 6.2: Transfer characteristic measurement for the pristine and formed device($E_{Form} = -2.01\frac{MV}{cm}$), as well as measurements during the temperature slope of an additional forming step at $T = 50°C$ and $T = 100°C$.

cumulated charge, obtained by integrating the displacement current. However, in consideration of a 30mm^2 area[24], the accumulated charge of 1024nC exceeds the maximum amount of charge, obtainable by a simple geometric capacitor model of the PMMA / SiO$_2$ layer stack. For an applied forming field strength of -1.34$\frac{MV}{cm}$, the accumulated charges should correspond to a maximum charge value of $\approx$ 125nC. This value is approximated, using equation 6.1:

$$Q = C_{Diel}V_{Form} \qquad (6.1)$$

Here, C_{Diel} represents the capacitance of the PMMA / SiO$_2$ dual layer dielectric. The obtained charge of 1024nC corresponds to a capacitance of $C_{Diel} = 25nF$. By assuming a constant perimitivity of $\epsilon_r = 3.9$ in the entire dielectric layer, an effective thickness of 40nm can be approximated for the OFET insulator. This thickness strongly suggests a positive volume charge, stored in the bulk of the SiO$_2$/PMMA dielectric layer stack.

By considering the transfer characteristic of the charged OFET, after its third forming step (F_F=-2.68$\frac{MV}{cm}$) (6.1(a)), an increase in I_D can be detected for low gate voltages. Since the gate current, recorded in parallel to the discussed measurement, allows to exclude gate leakage as the origin of the drain current, the increase in I_D is ascribed to a unipolar electron current. The injection of electrons into pentacene out of Au electrodes, despite a pronounced injection barrier of $\approx$ 2.2eV, is again conjectured to be promoted by a charged dielectric interface. Such a scenario was already discussed for the injection of holes out of Ca electrodes into pentacene in the previous chapter. The ambipolar characteristic of the OFET, after its third forming step, can be confirmed by a measurement in the electron accumulation. The obtained transfer characteristic is illustrated in Figure 6.3(b). For low gate voltages, the expected significant hole current is observed, while for high gate as well as high drain

---

[24] Due to the good injection and transport properties of holes for the investigated device, the entire pentacene area of the sample is used for this approximation.

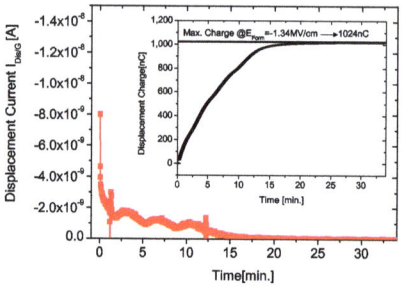 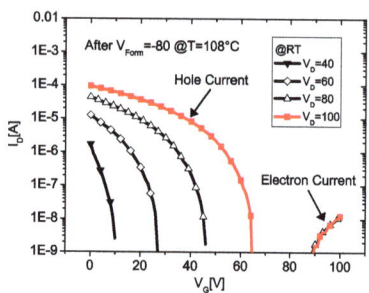

(a) Displacment current during the first PMMA forming step at $E_{Form} = -1.34 \frac{MV}{cm}$ and $T_{Form} = 108°C$. Inset: Corresponding charge, measured at the gate electrode

(b) Transfer characteristic in the electron accumulation mode, after the described 3rd forming step and the resulting ΔV_{th} =-31.3V.

Figure 6.3

voltages, an electron current can be detected. A threshold voltage for electrons of $V_{th,e} \approx 85$V as well as an electron mobility of $\mu_e \approx 1 * 10^{-4} \frac{cm^2}{Vs}$ is estimated. The electron injection out of Au electrodes into pentacene remained stable during the time frame of the measurement at room temperature (RT).

In the current chapter, resulting from forming steps at $T = 108°C$ and in dependence of the applied forming field strengths, a negative ΔV_{th} in the OFET transfer characteristics could be observed. The maximum observed shift in threshold voltage of $\Delta V_{th} = -31$V has allowed for a change in the OFET transport properties from unipolar p-type to ambipolar, and therefore an injection of electrons from the used Au metalization into pentacene. The observed charge carrier injection is possible, despite a difference in Au work function and pentacene electron affinity of about ≈ 2.2eV.

Chapter 7

Summary

Currently, for the realization of organic CMOS technology, two different types of organic semiconductors are required for the implementation of unipolar n- and p-type OFETs. In order to integrate these materials onto a single substrate, cost intensive processing steps are needed. Among other reasons, this represents one of the main hindrances for the application of organic CMOS technology in commercial products. In this thesis, the realization of OFETs with complementary polarity, using only a single organic semiconductor and even an identical device cross section was addressed by dielectric interface engineering. For OFET charge carrier transport the importance of electronic states at the dielectric / semiconductor interface is emphasized, which is an aspect that has been underestimated in the OFET development.

In order to investigate the influence of the dielectric / semiconductor interface on unipolar charge carrier transport by use of an ambipolar semiconductor, it was demonstrated, that the polarity of pentacene OFETs can be defined as unipolar n- or p-type by matching the metal work function of the source-drain metalization to either the electron affinity or the ionization potential of the organic semiconductor. With this approach, the influence of different polymers such as PS, PC, PMMA, P4VP and PI as gate insulators on the unipolar electron and hole charge carrier transport was investigated. A degradation in the n-type transistor performance with a decrease in water contact angle of the polymer dielectric interface due to an increase in oxygen containing polar groups was obtained. The experimental results suggest a strong influence of oxygen containing polar groups, such as keto and hydroxyl groups, at this interface on the electron charge carrier transport properties in pentacene OFETs. This is attributed to the electron trapping nature of these groups. At the same time, no hole trapping of keto and hydroxyl groups was observed. Using these results, two techniques were investigated in the following to selectively influence OFET charge carrier transport properties through the electron trapping in keto and hydroxyl groups.

In a first approach, electron transport in pentacene OFETs incorporating an SiO_2 dielectric was realized, despite a large number of electron traps (density -OH $\approx$ 10^{13}cm^{-2}) at the dielectric semiconductor interface. This was achieved by the deposition of a thin layer of Ca on the SiO_2 insulator, as previously demonstrated by Ahles et al. [83]. The investigation in this thesis has shown that the electron transport properties can be significantly improved with increasing Ca layer thickness, up to a value of d_{Ca} = 12Å. However, for layers exceeding this thickness the electron transport is degraded, until it is almost completely attenuated for Ca layer thicknesses exceeding 26Å. A detailed PES analysis of the Ca modified SiO_2 interface revealed the formation of an oxidized Ca layer on top of the dielectric surface, which in turn isolated and compensated available electron traps at the dielectric interface and thus enabling the n-type OFET charge carrier transport. The degradation in n-type performance for increasing Ca layer thicknesses could be linked to the formation of a metallic Ca fraction in the oxidized Ca layer. This was proposed to partially screen the electric field in the transistor channel, thereby negatively influencing the OFET charge carrier transport. The formation of the oxidized Ca layer was determined to be the result of an interface reaction between metallic Ca and substrate oxygen components. By enhancing the oxidizing reaction, using heat or an electrical cyclic stress, the OFET performance was significantly improved. This technique can be used to realize O-CMOS inverter stages, as shown by Ahles et al. [83].

In a second approach to influence the OFET charge carrier transport properties by dielectric interface engineering, electron traps in the form of keto and hydroxyl groups were introduced to the PMMA dielectric of an otherwise n-type pentacene OFET. These functional groups were obtained by exposing the polymer insulator to UV radiation in ambient atmosphere prior to the pentacene deposition. After negatively charging these traps, using an electrical cyclic conditioning in the electron accumulation, the polarity of the unipolar n-type OFET was changed to unipolar p-type without any other device modification. The change in polarity was due to a large threshold voltage shift of $\Delta V_{th} \approx 60V$ for electrons and holes. The injection of holes from Ca electrodes, despite a pronounced injection barrier, was proposed to occur as the result of an enhanced electric field in the transistor channel. This is due to trapped charges at the dielectric interface, resulting in a field enhanced charge carrier injection. This technique has enabled the realization of complementary unipolar n- and p-type pentacene OFETs with an identical device structure, exhibiting balanced charge carrier transport properties. It has furthermore allowed for the realization of an O-CMOS inverter stage with stable operation below its supply voltage and a high gain of 17.

It has been demonstrated, that the injection of holes from Ca electrodes into pentacene is possible despite a pronounced injection barrier of $\approx 2.1 eV$. In order to confirm, that the injection of complementary charge carriers into pentacene is also possible using a metalization stable in ambient atmosphere, such as Au, the electret characteristics of a PMMA dielectric were utilized. By positively charging the PMMA dielectric incorporated in a unipolar p-type OFET, a negative threshold voltage shift of $\Delta V_{th} \approx -31V$ was obtained. This was achieved by exposing the electret to high forming-fields at the glass temperature of PMMA. The shift in threshold leads to a change in the transistor polarity from unipolar p-type to ambipolar, thus substantiating the general possibility for the field enhanced injection of electrons from Au contacts into pentacene, despite a pronounced electron injection barrier of about 2.3eV.

Nomenclature

β	Transferintrgral
χ	Electron affinity
χ_{Pent}	Pentacene electron affinity
ΔE_d	Spectrometer resolution
ΔE_n	Line width of the of the excited photo hole E^f
ΔE_p	Line width of the excitation radiation
ΔE_{vac}	Vacuum level shift
ΔR_{ij}	Norm distance between two DOS states
$\Delta V_{th,e}$	Difference in electron threshold voltage
γ	Exponential wave function decay
$\lambda_m(E)$	Mean free path length
$\Delta E_{B2p\frac{3}{2}-2p\frac{1}{2}}$	Split in binding energy for a Ca2p dublett
$t_{exposure}$	UV exposure time
μ	Charge carrier mobility
μ_e	Electron mobility
μ_h	Hole mobility
Φ_{Met}	Metal work function
ϕ_{spec}	Work function spectrometer
ϕ_s	Sample work function
σ	Conductivity
σ	Standard deviation

τ	Mean free time of a charge carrier
$\Theta(x)$	Heaviside step function
a_0	Distance between two molecules
$alpha$	Angle between photon path and detected electron path
C_{Diel}	Capacitance of the transistor or MIS diode dielectric
C_{Tot}	Total device capacitance of the dielectric as well as the semiconductor
C_{tot}	Total device capacitance
d_{eff}	Effective thickness
E	Energy
e	Elemental charge
E_{Bind}	Binding energy
E_C	Activation energy for conduction
E_{Form}	Forming Field
E_F	Fermi Level
E_F	Trap state energy level
E_i	Initial energetic DOS state
E_j	Target energetic DOS state
E_{kin}	Kinetic Energy
I	Ionisation potential
$I_{D,max}$	Maximum drain current
j	Total momentum
k	Boltzmann constant
l	Angular momentum
n	Total amount of accumulated charge
$n(x)$	Electron density per unit area at position x
$n_{th,e}$	Threshold charge carrier density for electrons

Nomenclature

$n_{th,h}$	Threshold charge carrier density for holes
n_{th}	Threshold charge carrier density
n_t	trapped electrons
$p(x)$	Electron density per unit area at position x
$P_{e/h}$	Electronic polarization energy for electrons / hole
$Q(x)$	Net surface charge
$R(x_0)$	Channel resistance integrated up to a random position x_0 of the transistor channel
$R'(x)$	Partial channel resistance
s	Spin momentum
S_C	atomic sensitivity factor for carbon
S_O	atomic sensitivity factor for oxygen
T	Temperature
T_F	Forming Temperature
T_n	Lifetime of an excited hole state
$V(x)$	Voltage drop along the transistor dielectric surface
V_{DD}	Inverter supply voltage
V_{In}	Inverter input voltage
$V_{th,e}$	Threshold voltage for electron conduction
$V_{th,h}$	Threshold voltage for hole conduction
AC	Alternating Current
AFM	Atomic force microscopy
ASF	Atomic sensitivity factor
C	Capacitance per unit area
CA	Contact Angle
CMOS	Complementary Metal Oxide Semiconductor
DAISY-MAT	Darmstadt Integrated System for Materials Science

DC	Direct current
DOS	Density Of States
E	Electric field strength
e	Elemental charge
HOMO	Highest Occupied Molecular Orbital
LUMO	Lowest Unoccupied Molecular Orbital
MIS	Metal insulator semiconductor
O-CMOS	Organic complementary metal oxide semiconductor
OLED	Organic Light Emitting Diode
PES	Photoelectron spectroscopy
PMMA	Polymethylmethacryl
PVD	Physical Vapor Deposition
PVD	Physical vapor deposition
RT	Room Temperature
S	Peak area atmoic sensitivity factor
SAM	Self assembled monolayer
TFT	Thin Film Transistor
THF	Tetra Hydro Furan
TSC	Thermally stimulated discharge
UV	Ultra violet
XPS	X-Ray photoelectron spectroscopy

List of Figures

1.1 Development in OFET p-type mobility between 1984 and 2007. The various p-type materials are grouped together in families of similar molecular core part. The shaded bars represent the a-Si:H and Poly-Si mobility range [11]. 4

2.1 Pentacene energy diagram with respect to the electron affinity and the ionisation potential. The values are depicted for the gas phase 1) (χ_G, I_G), as well as the single crystal form 2) (χ_C, I_C) of pentacene. [31] 8

2.2 Transport states in a disordered molecular solid. Three different hopping transitions a), b) and c) are illustrated by the inset. Only for transition c), an activation energy is required prior to the tunneling process. 10

2.3 Hydroxyl group electron trapping mechanism suggest by Chua et al. [29]. 12

2.4 Standard OFET designs in top / bottom source-drain contact configuration. The top / bottom gate architecture indicated in Figures 2.4(a) and 2.4(b), can be applied to both of the source-drain contact configurations. 13

2.5 Schematic illustration of typical transistor current voltage characteristics. For $V_D \leq V_G$ the transistor is always operated in the unipolar range (dotted lines). For $V_D > V_G$ uni- or ambipolar (dotted lines) operation is possible. OFETs that allow for the injection and transport of electrons as well as holes exhibit ambipolar operation in that voltage range. Otherwise, the transistor characteristic saturates and remains unipolar. 14

2.6 Resistor capacitor equivalent circuit for the extended Schockley transistor model [33]. 15

2.7 Charge carriers contributing to the ambipolar current for $|V_D| > |V_G|$ 16

2.8 Impedance and DC measurements on MIS diodes consisting of a p^{++}-Si / insulator / pentacene / Ca layer stack [54] 19

2.9 Electret charging by contacting electrode. 1) Dipole alignment and space charge separation. 2) Charge carrier injection due to high electrical field strenghts. 20

3.1 Chemical structure of the polymers used as dielectric materials. The abbreviations used for the individual polymers are summarized in Table 3.1. 23

3.2 25

3.3 Schematic illustration of the used top contact transistor structure geometry. 27

3.4 Driving schematic used for the different electrical cyclic conditioning steps. 28

3.5 PMMA dielectric forming setup and equivalent circuit. 29

3.6 Thermal forming step experimental setup, excluding the HP4155A parameter analyzer. 29

3.7 Schematic energy level diagram during the photoelectron excitation and detection [78]. The binding energy is measured with respect to the work function of the spectrometer. A photoelectron has to surpass the energy difference between the sample and spectrometer work functions ($\Phi_{spec} - \Phi_s$). 32

4.1 Development in OFET pentacene n / h mobility between 1992 and 2007. The shaded bar represent the a-Si:H mobility range [11]. 37

4.2 Work function of Au and Ca relative to pentacene HOMO / LUMO levels. 38

4.3 Transfer characteristics of pentacene OFETs, differing only in the source-drain metalization. By matching the metal work function to the respective pentacene HOMO / LUMO levels, the transistor polarity is defined as either unipolar n- / p-type or ambipolar. . . . 38

4.4 Unipolar output characteristics in the hole and electron accumulation. The filled and open scatter plots represent the respective unipolar p-type and n-type characteristic of a Au / Au or Ca / Ca electrode pentacene OFET. 39

4.5 n_{th} and μ in dependence of the water contact angle. For the error bar statistic up to 4 samples out of different batches were utilized, with a maximum of 3 OFETs per sample. 41

5.1 Output and transfer characteristic of a pentacene OFET incorporating Ca source-drain contacts as well as a SiO_2 insulator comprising a 8Å Ca layer at the interface. 44

5.2 a) Pentacene OFET electron mobility as well as b) electron threshold voltage and $\frac{On}{Off}$ ratio, as extracted from the electron accumulation in dependence of the Ca interlayer thickness. 45

5.3 XP-survey spectra on SiO_2 for ascending Ca layer thickness. 46

5.4 Ca2p emission spectra on SiO_2 for ascending Ca layer thickness. The Ca layer thickness is enhanced in between the respective measurements without breaking the vacuum. The spectra have been shifted in binding energy relative to the metallic Ca2p emission at an adsorbate thickness of 235Å for better comparability. The as measured binding energy values are illustrated in Figure 5.7. 47

5.5 Ca2p emission spectrum of only the oxidized component for an adsorbate thickness of d=53Å. The metallic component has been subtracted using the scaled Ca2p emission at d=235Å. The fit exhibiting the closed triangle plot represents the -CaO component, while the closed square plot represents the -CaOH component. The residuum of the fit is given at the bottom of the graph. 48

5.6 O1s and Si2p emission spectra. The spectra have been shifted in binding energy with respect to the substrate emission line obtained for a Ca thickness of 0Å, for better comparability. The as measured binding energy values are illustrated in Figure 5.7. The as measured spectra are illustrated in Appendix B. 49

5.7 As measured binding energies for the Ca2p, Si2p and O1s emission lines, in dependence of the Ca adsorbate thickness. 50

List of Figures

5.8 O1s emission spectra for an unheated / heated Si p^{++}-Si substrate with a 200nm thermally grown dry oxide comprising a Ca adsorbate (d_{Ca} = 50Å). The temper step was conducted in the UHV. Fits to the individual components of the spectra as well as their error function at the bottom of each spectrum are illustrated by both Figures. The 5 times magnified residuum indicates the deviation of the fit to the measurement. 52

5.9 Electron mobility and threshold voltage (at RT) for OFETs comprising a d_{Ca} = 8Å Ca passivation layer, in dependence of different annealing temperatures during the production process. The annealing was conducted in inert N$_2$ atmosphere for the duration of t=1h. 53

5.10 Improvement in I_D for an n-type OFET comprising a 8Å passsivated SiO$_2$ dielectric, during the described electrical cyclic conditioning step. 55

5.11 Comparison of the output characteristics for pentacene OFETs comprising a 8Å Ca interlayer in their pristine, annealed (t = 1h at T_a = 160°C) and electrically cyclic conditioned state. Furthermore the characteristic for an OFET comprising a pristine SiO$_2$ / pentacene interface is illustrated. 56

5.12 OFET charge carrier transport parameters in dependence of the Ca passivation thickness. 57

5.13 Comparison in PMMA topography for a pristine and UV irradiated layer deposited on SiO$_2$. 58

5.14 Water contact angle for PMMA in dependence of the UV exposure time. 59

5.15 Area normalized O1s and C1s emission spectra of PMMA and UV modified PMMA. 60

5.16 PMMA chemical structure O=C is marked by a), and O-C is marked by b) 60

5.17 O1s PMMA emission lines, recorded using Mg Kα radiation at a sample / analyzer angle of 45°. The error graph at the bottom of each figure indicates the deviation between experimental data and the conducted fit. 61

5.18 Reaction mechanism in the PMMA near surface layer leading to the formation of hydroxyl groups during UV irradiation, as suggested by Wei et al. [61]. 62

5.19 AFM micrographs of pentacene deposited onto different dielectric layers, obtained in the non contact mode. 63

5.20 Pentacene profile of the profile lines indicated in Figure 5.19(c) 64

5.21 Output characteristic for a pentacene OFET comprising a UV modified PMMA dielectric in its pristine and electrical cyclic conditioned state. 66

5.22 Threshold voltage shift between the 1st and 8th electrical cyclic conditioning step for an OFET incoperating a UV modified PMMA gate dielectric. 66

5.23 Field enhanced tunnel injection. 68

5.24 Degradation in $I_{D,max}$ for a UV-modified p-type OFET over time. The squares represent values derived from output characteristic measurements. The vertical dotted lines indicate the measurement of a transfer characteristic. All measurements were conducted in the hole accumulation. 70

5.25 Degradation as well as reconditioning of a UV modified OFET in the UHV (<1 ∗ 10^{-10}mbar), depicted for the case of output characteristics recorded at V_G = 80V. ... 71

5.26	Transfer characteristic of complementary OFETs. In analogy to an inverter structure, the source contact of the p-type OFET is held at a potential of 60V for both plots.	73
5.27	Organic CMOS inverter stage transfer characteristic.	74
6.1	Influence of the respective forming steps on the OFET device characteristic.	78
6.2	Transfer characteristic measurement for the pristine and formed device(E_{Form} = -2.01 $\frac{MV}{cm}$), as well as measurements during the temperature slope of an additional forming step at $T = 50°C$ and $T = 100°C$.	79
6.3		80
B.1	Ca2p emission spectra on SiO_2 for ascending Ca layer thickness. The Ca layer thickness is enhanced in between the respective measurements without breaking the vacuum.	106
B.2	O1s emission spectra for ascending Ca layer thicknesson SiO_2. The Ca layer thickness is enhanced in between the respective measurements without breaking the vacuum.	107
B.3	Si2p emission spectra for ascending Ca layer thickness on SiO_2. The Ca layer thickness is enhanced in between the respective measurements without breaking the vacuum.	108

List of Tables

3.1	Specifications of the utilized insulating polymers.	24
3.2	Summary of the evaporated materials and their corresponding crucibles. The sublimation temperature of the materials is specified for a pressure of 10^{-2} mbar.	26
3.3	Angular (l), spin (s) and total momentum (j) of the atomic s- and p-orbitals. In addition, the intensity ratio ($\frac{2j+1}{2j'+1}$) for possible dublett states of photoelectrons is indicated.	33
3.4	Specifications of the analyzer unit PHI5700 (Physical Electronics).	34
5.1	Difference in binding energy between Ca and -CaO, -CaOH in a CaNi5 alloy. Resulting Ca2p -CaO and -CaOH binding energies for Ca on SiO_2.	48
5.2	Oxygen components for a $p^{++} - Si / SiO_2$(200nm) substrate comprising a $d_{Ca} = 8\text{Å}$ Ca layer in its unheated and annealed state. The substrate was annealed at $T = 180°C$ for the duration of $t = 1h$.	53
5.3	p-type pentacene OFET charge carrier transport parameters at RT. The values were extracted prior to and after an annealing step at $T=160°C$ for the duration of $t=1h$.	54
5.4	Comparison in OFET charge carrier transport parameters for devices containing a 8Å Ca passivation layer in their pristine, electrical cyclic conditioned and annealed state.	55
5.5	Fit parameters for the deconvoluted O1s emission spectra, as well as the carbon and oxygen concentration for PMMA after and prior to the UV irradiation for 10min. in ambient atmosphere.	61
5.6	Summary of the data extracted from the interface analysis of SiO_2, PMMA and Pentacene, as discussed in section 5.2.1.	64
5.7	Thermal activation required for tunnel injection through 1.5nm and 4nm distances at different field strengths an pentacene layer thicknesses.	69
5.8	Degradation in OFET threshold and mobility over time.	71
C.1	Single crystal OFET p-type mobility between 1992-2007(Figure 1.1).	109
C.2	Pentacene OFET mobility between 1992-2007 (Figure 4.1).	109
C.3	Development in OFET p-type mobility between 1984 and 2007. For the mobility development of figure 1.1, the various p-type materials are grouped together in families of similar molecular core parts.	110

D.1 Process parameters for the polymer deposition in inert N_2. The exact dilutions of the respective polymers are listed in table D.2 111
D.2 Abbreviations and dilutions of the utilized polymers 112
D.3 Process parameters for evaporated thin films 112

Literature

[1] A. Pochettino, *Acad. Lincei. Rediconti.*, vol. 15, p. 355, 1906.

[2] M. Volmer, *Ann. Phys.*, vol. 40, p. 775, 1913.

[3] M. Pope and C. E. Swenberg, *Electronic Processes in Ogrganic Crystals.* Clarndon Press, 1982.

[4] H. Hoegl, O. Sues, and W. Neugebauer, *1068115*, German Patent, 1957.

[5] H. G. Greig, "An organic photoconductive system," *Rca Review*, vol. 23, no. 3, pp. 413–419, 1962.

[6] A. Bernanose, M. Comte, and P. Vouaux, "*sur un nouveau mode demission lumineuse chez certains composes organiques," *Journal De Chimie Physique Et De Physico-Chimie Biologique*, vol. 50, no. 1, pp. 64–68, 1953.

[7] W. Helfrich and W. G. Schneide, "Recombination radiation in anthracene crystals," *Physical Review Letters*, vol. 14, no. 7, pp. 229–&, 1965.

[8] D. F. Barbe and C. R. Westgate, "Surface state parameters of metal-free phthalocyanine single crystals," *Journal Of Physics And Chemistry Of Solids*, vol. 31, no. 12, pp. 2679–&, 1970.

[9] K. Kudo, M. Yamashina, and T. Moriizumi, "Field-effect measurement of organic-dye films," *Japanese Journal Of Applied Physics Part 1-Regular Papers Short Notes & Review Papers*, vol. 23, no. 1, pp. 130–130, 1984.

[10] A. Tsumura, H. Koezuka, and T. Ando, "Macromolecular electronic device - field-effect transistor with a polythiophene thin-film," *Applied Physics Letters*, vol. 49, no. 18, pp. 1210–1212, Nov. 1986.

[11] T. K. Chuang, M. Troccoli, M. Hatalis, and A. T. Voutsas, "Polysilicon tft technology on flexible metal foil for ampled displays," *Journal Of The Society For Information Display*, vol. 15, no. 7, pp. 455–461, 2007.

[12] K. Shankar and T. N. Jackson, "Morphology and electrical transport in pentacene films on silylated oxide surfaces," *Journal Of Materials Research*, vol. 19, no. 7, pp. 2003–2007, 2004.

[13] Y. Y. Lin, D. J. Gundlach, S. F. Nelson, and T. N. Jackson, "Stacked pentacene layer organic thin-film transistors with improved characteristics," *Ieee Electron Device Letters*, vol. 18, no. 12, pp. 606–608, Dec. 1997.

[14] H. Sirringhaus, P. J. Brown, R. H. Friend, M. M. Nielsen, K. Bechgaard, B. M. W. Langeveld-Voss, A. J. H. Spiering, R. A. J. Janssen, E. W. Meijer, P. Herwig, and D. M. de Leeuw, "Two-dimensional charge transport in self-organized, high-mobility conjugated polymers," *Nature*, vol. 401, no. 6754, pp. 685–688, Oct. 1999.

[15] A. Salleo, M. L. Chabinyc, M. S. Yang, and R. A. Street, "Polymer thin-film transistors with chemically modified dielectric interfaces," *Applied Physics Letters*, vol. 81, no. 23, pp. 4383–4385, Dec. 2002.

[16] D. Knipp, R. A. Street, A. Volkel, and J. Ho, "Pentacene thin film transistors on inorganic dielectrics: Morphology, structural properties, and electronic transport," *Journal Of Applied Physics*, vol. 93, no. 1, pp. 347–355, Jan. 2003.

[17] M. M. Ling, P. Erk, M. Gomez, M. Koenemann, J. Locklin, and Z. N. Bao, "Air-stable n-channel organic semiconductors based on perylene diimide derivatives without strong electron withdrawing groups," *Advanced Materials*, vol. 19, no. 8, pp. 1123–1127, Apr. 2007.

[18] S. Janietz, J. Barche, A. Wedel, and D. Sainova, "n-type copolymers with fluorene and 1,3,4-heterodiazole moieties," *Macromolecular Chemistry And Physics*, vol. 205, no. 2, pp. 187–198, Jan. 2004.

[19] J. Zaumseil and H. Sirringhaus, "Electron and ambipolar transport in organic field-effect transistors," *Chemical Reviews*, vol. 107, no. 4, pp. 1296–1323, Apr. 2007.

[20] Y. Sakamoto, T. Suzuki, M. Kobayashi, Y. Gao, Y. Fukai, Y. Inoue, F. Sato, and S. Tokito, "Perfluoropentacene: High-performance p-n junctions and complementary circuits with pentacene," *Journal Of The American Chemical Society*, vol. 126, no. 26, pp. 8138–8140, Jul. 2004.

[21] S. De Vusser, S. Steudel, K. Myny, J. Genoe, and P. Heremans, "Low voltage complementary organic inverters," *Applied Physics Letters*, vol. 88, no. 16, p. 162116, Apr. 2006.

[22] M. Kitamura and Y. Arakawa, "Low-voltage-operating complementary inverters with c60 and pentacene transistors on glass substrates," *Applied Physics Letters*, vol. 91, no. 5, p. 053505, 2007.

[23] M. M. Ling, Z. N. Bao, P. Erk, M. Koenemann, and M. Gomez, "Complementary inverter using high mobility air-stable perylene di-imide derivatives," *Applied Physics Letters*, vol. 90, no. 9, p. 093508, Feb. 2007.

[24] H. Klauk, U. Zschieschang, J. Pflaum, and M. Halik, "Ultralow-power organic complementary circuits," *Nature*, vol. 445, no. 7129, pp. 745–748, Feb. 2007.

[25] T. Kawase, T. Shimoda, C. Newsome, H. Sirringhaus, and R. H. Friend, "Inkjet printing of polymer thin film transistors," *Thin Solid Films*, vol. 438, pp. 279–287, Aug. 2003.

[26] A. Knobloch, A. Manuelli, A. Bernds, and W. Clemens, "Fully printed integrated circuits from solution processable polymers," *Journal Of Applied Physics*, vol. 96, no. 4, pp. 2286–2291, Aug. 2004.

[27] V. Subramanian, P. C. Chang, J. B. Lee, S. E. Molesa, and S. K. Volkman, "Printed organic transistors for ultra-low-cost rfid applications," *Ieee Transactions On Components And Packaging Technologies*, vol. 28, no. 4, pp. 742–747, Dec. 2005.

[28] M. Ahles, R. Schmechel, and H. von Seggern, "n-type organic field-effect transistor based on interface-doped pentacene," *Applied Physics Letters*, vol. 85, no. 19, pp. 4499–4501, Nov. 2004.

[29] L. L. Chua, J. Zaumseil, J. F. Chang, E. C. W. Ou, P. K. H. Ho, H. Sirringhaus, and R. H. Friend, "General observation of n-type field-effect behaviour in organic semiconductors," *Nature*, vol. 434, no. 7030, pp. 194–199, Mar. 2005.

[30] G. M. Sessler, J. van Turnhout, B. Gross, M. G. Broadhurst, G. T. Davis, S. Mascarenhas, J. E. West, and R. Gerhard-Multhaupt, *Topics in Applied Physics*, 2nd ed., ser. Electrets, G. M. Sessler, Ed. Springer, 1987, vol. 33.

[31] M. Schwoerer and H. C. Wolf, *Organische Molekulare Festkoerperphysik*. Wiley-VCH, 2005.

[32] W. Bruetting, Ed., *Physics of Organic Semiconductors*. Wiley-VCH, 2005.

[33] R. Schmechel, M. Ahles, and H. von Seggern, "A pentacene ambipolar transistor: Experiment and theory," *Journal Of Applied Physics*, vol. 98, no. 8, p. 084511, Oct. 2005.

[34] W. Warta, R. Stehle, and N. Karl, "Ultrapure, high mobility organic photoconductors," *Applied Physics A-Materials Science & Processing*, vol. 36, no. 3, pp. 163–170, 1985.

[35] Y. C. Cheng, R. J. Silbey, D. A. da Silva, J. P. Calbert, J. Cornil, and J. L. Bredas, "Three-dimensional band structure and bandlike mobility in oligoacene single crystals: A theoretical investigation," *Journal Of Chemical Physics*, vol. 118, no. 8, pp. 3764–3774, Feb. 2003.

[36] E. Silinsh and C. Vladislav, *Oganic Molecular Crystals*. American Institute of Physics, 1994.

[37] R. Schmechel, "Manuscript to organic semiconductors ws 2004," TU Darmstadt, Materials Science Departement, Tech. Rep.

[38] P. M. Borsenberger, L. Pautmeier, and H. Bassler, "Hole transport in bis(4-n,n-diethylamino-2-methylphenyl)-4-methylphenylmethane," *Journal Of Chemical Physics*, vol. 95, no. 2, pp. 1258–1265, Jul. 1991.

[39] H. Baessler, "Charge transport in disordered organic photoconductors - a monte-carlo simulation study," *Physica Status Solidi B-Basic Research*, vol. 175, no. 1, pp. 15–56, Jan. 1993.

[40] R. Schmechel, "Hopping transport in doped organic semiconductors: A theoretical approach and its application to p-doped zinc-phthalocyanine," *Journal Of Applied Physics*, vol. 93, no. 8, pp. 4653–4660, Apr. 2003.

[41] A. Miller and E. Abrahams, "Impurity conduction at low concentrations," *Physical Review*, vol. 120, no. 3, pp. 745–755, 1960.

[42] A. Lampert, M. and P. Mark, *Current Injection in Solids*. Academic Press, 1970.

[43] A. Kadashchuk, R. Schmechel, H. von Seggern, U. Scherf, and A. Vakhnin, "Charge-carrier trapping in polyfluorene-type conjugated polymers," *Journal Of Applied Physics*, vol. 98, no. 2, p. 024101, Jul. 2005.

[44] N. Benson, C. Melzer, R. Schmechel, and H. von Seggern, "Electronic states at the dielectric/semiconductor interface in organic field effect transistors," *physica status solidi (a)*, vol. 205, no. 3, pp. 475–487, 2008.

[45] N. von Malm, "Ladungstraegerfallen in amorphen organischen halbleitern," Ph.D. dissertation, Technische Universitaet Darmstadt, Fachbereich Material- und Geowissenschaften, Fachgebiet Elektronische Materialeigenschaften, 2003.

[46] W. Shockley, M. Sparks, and G. K. Teal, "P-n junction transistors," *Physical Review*, vol. 83, no. 1, pp. 151–162, 1951.

[47] J. E. Lilienfeld, "U.s. patent 1.745.175 and u.s. patent 1.900.018 and u.s. patent 1.877.140."

[48] T. Lindner and G. Paasch, "Inversion layer formation in organic field-effect devices," *Journal Of Applied Physics*, vol. 102, no. 5, p. 054514, Sep. 2007.

[49] W. Shockley, "A unipolar field-effect transistor," *Proceedings Of The Institute Of Radio Engineers*, vol. 40, no. 11, pp. 1365–1376, 1952.

[50] J. Veres, S. Ogier, G. Lloyd, and D. de Leeuw, "Gate insulators in organic field-effect transistors," *Chemistry Of Materials*, vol. 16, no. 23, pp. 4543–4555, Nov. 2004.

[51] M. Halik, H. Klauk, U. Zschieschang, G. Schmid, C. Dehm, M. Schutz, S. Maisch, F. Effenberger, M. Brunnbauer, and F. Stellacci, "Low-voltage organic transistors with an amorphous molecular gate dielectric," *Nature*, vol. 431, no. 7011, pp. 963–966, Oct. 2004.

[52] J. Veres, S. D. Ogier, S. W. Leeming, D. C. Cupertino, and S. M. Khaffaf, "Low-k insulators as the choice of dielectrics in organic field-effect transistors," *Advanced Functional Materials*, vol. 13, no. 3, pp. 199–204, Mar. 2003.

[53] I. N. Hulea, S. Fratini, H. Xie, C. L. Mulder, N. N. Iossad, G. Rastelli, S. Ciuchi, and A. F. Morpurgo, "Tunable frohlich polarons in organic single-crystal transistors," *Nature Materials*, vol. 5, no. 12, pp. 982–986, Dec. 2006.

[54] M. Ahles, "Einfluss der dotierung organischer halbleiter auf den feldeffekt," Ph.D. dissertation, Technische Universitaet Darmstadt, Fachbereich Material- und Geowissenschaften, Fachgebiet Elektronische Materialeigenschaften, 2005.

[55] E. J. Meijer, D. M. De Leeuw, S. Setayesh, E. Van Veenendaal, B. H. Huisman, P. W. M. Blom, J. C. Hummelen, U. Scherf, and T. M. Klapwijk, "Solution-processed ambipolar organic field-effect transistors and inverters," *Nature Materials*, vol. 2, no. 10, pp. 678–682, Oct. 2003.

[56] J. van Turnhout, *Thermally Stimulated Discharge of Polymer Electrets.* Elsevier, 1975.

[57] R. G. Vyverberg, *Charging Photoconduction Surfaces*, J. H. Dessauer and E. Clark, H., Eds. Focal Press, London, 1965.

[58] H. Stoecker, *Taschenbuch der Physik.* Harri Deutsch, Frankfurt am Main, 3te Auflage, 1998.

[59] A. Hozumi, H. Inagaki, and T. Kameyama, "The hydrophilization of polystyrene substrates by 172-nm vacuum ultraviolet light," *Journal Of Colloid And Interface Science*, vol. 278, no. 2, pp. 383–392, Oct. 2004.

[60] Y. J. Liu, D. Ganser, A. Schneider, R. Liu, P. Grodzinski, and N. Kroutchinina, "Microfabricated polycarbonate ce devices for dna analysis," *Analytical Chemistry*, vol. 73, no. 17, pp. 4196–4201, Sep. 2001.

[61] S. Y. Wei, B. Vaidya, A. B. Patel, S. A. Soper, and R. L. McCarley, "Photochemically patterned poly(methyl methacrylate) surfaces used in the fabrication of microanalytical devices," *Journal Of Physical Chemistry B*, vol. 109, no. 35, pp. 16988–16996, Sep. 2005.

[62] J. N. Chai, F. Z. Lu, B. M. Li, and D. Y. Kwok, "Wettability interpretation of oxygen plasma modified poly(methyl methacrylate)," *Langmuir*, vol. 20, no. 25, pp. 10919–10927, Dec. 2004.

[63] A. Hozumi, T. Masuda, K. Hayashi, H. Sugimura, O. Takai, and T. Kameyama, "Spatially defined surface modification of poly(methyl methacrylate) using 172 nm vacuum ultraviolet light," *Langmuir*, vol. 18, no. 23, pp. 9022–9027, Nov. 2002.

[64] http://www2.dupont.com/Kapton, "Kapton polyimide fim general specifications," Bulletin Gs-96-7, 2008.

[65] M. Naddaf, C. Balasubramanian, P. S. Alegaonkar, V. N. Bhoraskar, A. B. Mandle, V. Ganeshan, and S. V. Bhoraskar, "Surface interaction of polyimide with oxygen ecr plasma," *Nuclear Instruments & Methods In Physics Research Section B-Beam Interactions With Materials And Atoms*, vol. 222, no. 1-2, pp. 135–144, Jul. 2004.

[66] R. Ruiz, D. Choudhary, B. Nickel, T. Toccoli, K. C. Chang, A. C. Mayer, P. Clancy, J. M. Blakely, R. L. Headrick, S. Iannotta, and G. G. Malliaras, "Pentacene thin film growth," *Chemistry Of Materials*, vol. 16, no. 23, pp. 4497–4508, Nov. 2004.

[67] T. Yasuda, T. Goto, K. Fujita, and T. Tsutsui, "Ambipolar pentacene field-effect transistors with calcium source-drain electrodes," *Applied Physics Letters*, vol. 85, no. 11, pp. 2098–2100, Sep. 2004.

[68] H. Heil, "Injektion, transport und elektrolumineszenz in organischen halbleiterbauelementen," Ph.D. dissertation, Technische Universitaet Darmstadt, Fachbereich Material- und Geowissenschaften, Fachgebiet Elektronische Materialeigenschaften, 2004.

[69] A. Hepp, "Der leuchtende organische feldeffekttransistor," Ph.D. dissertation, Technische Universitaet Darmstadt, Fachbereich Material- und Geowissenschaften, Fachgebiet Elektronische Materialeigenschaften, 2005.

[70] Landolt and Boernstein, *Vapor Pressure of Chemicals, Subvolume A, Volume 20*. Springer, 2001.

[71] *Benutzerhandbuch G10 Kontaktwinkelmessgeraet*. Kruess GmbH, Hamburg, 1994-1995.

[72] C. M. Chan, *Polymer Surface Modification and Characterization*. Hanser, 1994.

[73] M. Morra, E. Occhiello, and F. Garbassi, "Knowledge about polymer surfaces from contact-angle measurements," *Advances In Colloid And Interface Science*, vol. 32, no. 1, pp. 79–116, Jun. 1990.

[74] H. Ibach and H. Lueth, *Festkoerperphysik*. Springer, 1999.

[75] S. Huefner, *Photoelectron Spectroscopy*, K. V. Lotsch, H., Ed. Springer, 1995.

[76] J. F. Moulder, W. F. Stickle, P. E. Sobol, and D. Bomben, K., *Handbook of X-ray Photoelectron Spectroscopy*, J. Chastain and C. J. King, R., Eds. Physical Electronics, Inc., 1995.

[77] D. Ensling, "Photoelektronenspektroskopische untersuchung der elektronischen struktur duenner lithiumkobaltoxidschichten," Ph.D. dissertation, Technische Universitaet Darmstadt, Fachbereich Material- und Geowissenschaften, Fachgebiet Elektronische Materialeigenschaften, 2006.

[78] Y. Gassenbauer and A. Thissen, *Grundlagen der Roentgenphotoelektronenspektroskopie (X-Ray Photoelectron Spectroscopy "XPS")*, Technische Universitaet Darmstadt, FB Material- und Geowissenschaften, FG Oberflaechenforschung, 2006. [Online]. Available: http://www.tu-darmstadt.de/surface

[79] W. Atkins, P., *Physikalische Chemie*, A. Hoepfner, Ed. Wiley-VCH, 2001.

[80] [Online]. Available: http://srdata.nist.gov/xps/Default.aspx

[81] D. A. Shirley, "High-resolution x-ray photoemission spectrum of valence bands of gold," *Physical Review B*, vol. 5, no. 12, pp. 4709–&, 1972.

[82] T. Minakata, M. Ozaki, and H. Imai, "Conducting thin-films of pentacene doped with alkaline-metals," *Journal Of Applied Physics*, vol. 74, no. 2, pp. 1079–1082, Jul. 1993.

[83] M. Ahles, R. Schmechel, and H. von Seggern, "Complementary inverter based on interface doped pentacene," *Applied Physics Letters*, vol. 87, no. 11, p. 113505, Sep. 2005.

[84] N. Benson, M. Schidleja, C. Siol, C. Melzer, and H. von Seggern, "Dielectric interface modification by uv irradiation: a novel method to control ofet charge carrier transport properties," *Spie Proceedings*, vol. 6658, p. 66580W, 2007.

[85] N. Karl, "Charge carrier transport in organic semiconductors," *Synthetic Metals*, vol. 133, pp. 649–657, Mar. 2003.

[86] H. van Doveren and J. A. T. Verhoeven, "Xps spectra of ca, sr, ba and their oxides," *Journal Of Electron Spectroscopy And Related Phenomena*, vol. 21, no. 3, pp. 265–273, 1980.

[87] P. Selvam, B. Viswanathan, and V. Srinivasan, "Xps studies of the surface-properties of cani5," *Journal Of Electron Spectroscopy And Related Phenomena*, vol. 49, no. 3, pp. 203–211, Sep. 1989.

[88] A. Torikai, M. Ohno, and K. Fueki, "Photodegradation of poly(methyl methacrylate) by monochromatic light - quantum yield, effect of wavelengths, and light-intensity," *Journal Of Applied Polymer Science*, vol. 41, no. 5-6, pp. 1023–1032, 1990.

[89] E. Breitmeier and G. Jung, *Organische Chemie*. Thieme, 2005.

[90] K. Shin, C. W. Yang, S. Y. Yang, H. Y. Jeon, and C. E. Park, "Effects of polymer gate dielectrics roughness on pentacene field-effect transistors," *Applied Physics Letters*, vol. 88, no. 7, p. 072109, Feb. 2006.

[91] I. Kymissis, C. D. Dimitrakopoulos, and S. Purushothaman, "High-performance bottom electrode organic thin-film transistors," *Ieee Transactions On Electron Devices*, vol. 48, no. 6, pp. 1060–1064, Jun. 2001.

[92] C. D. Dimitrakopoulos and P. R. L. Malenfant, "Organic thin film transistors for large area electronics," *Advanced Materials*, vol. 14, no. 2, pp. 99–+, Jan. 2002.

[93] N. Koch, "Organic electronic devices and their functional interfaces," *Chemphyschem*, vol. 8, no. 10, pp. 1438–1455, Jul. 2007.

[94] H. Ishii, K. Sugiyama, E. Ito, and K. Seki, "Energy level alignment and interfacial electronic structures at organic metal and organic organic interfaces," *Advanced Materials*, vol. 11, no. 8, pp. 605–+, Jun. 1999.

[95] N. J. Watkins, L. Yan, and Y. L. Gao, "Electronic structure symmetry of interfaces between pentacene and metals," *Applied Physics Letters*, vol. 80, no. 23, pp. 4384–4386, Jun. 2002.

[96] S. M. Sze and K. N. Kwok, *Physics of Semiconductor Devices*, 3rd ed. Wiley-Interscience, 2007.

[97] N. Benson, M. Schidleja, C. Melzer, R. Schmechel, and H. von Seggern, "Complementary organic field effect transistors by ultraviolet dielectric interface modification," *Applied Physics Letters*, vol. 89, no. 18, p. 182105, Oct. 2006.

[98] H. v. S. Marcus Ahles, Roland Schmechel, "Organic cmos technology based on interface doped pentacene," *Materials Research Society Symposium*, vol. 871E, pp. I4.10.1 – I4.10.6, 2005.

[99] V. Podzorov and M. E. Gershenson, "Photoinduced charge transfer across the interface between organic molecular crystals and polymers," *Physical Review Letters*, vol. 95, no. 1, p. 016602, Jul. 2005.

[100] K. J. Baeg, Y. Y. Noh, J. Ghim, S. J. Kang, H. Lee, and D. Y. Kim, "Organic non-volatile memory based on pentacene field-effect transistors using a polymeric gate electret," *Advanced Materials*, vol. 18, no. 23, p. 3179, Dec. 2006.

[101] V. Y. Butko, X. Chi, D. V. Lang, and A. P. Ramirez, "Field-effect transistor on pentacene single crystal," *Applied Physics Letters*, vol. 83, no. 23, pp. 4773–4775, Dec. 2003.

[102] V. Podzorov, S. E. Sysoev, E. Loginova, V. M. Pudalov, and M. E. Gershenson, "Single-crystal organic field effect transistors with the hole mobility similar to 8 cm(2)/v s," *Applied Physics Letters*, vol. 83, no. 17, pp. 3504–3506, Oct. 2003.

[103] C. Goldmann, S. Haas, C. Krellner, K. P. Pernstich, D. J. Gundlach, and B. Batlogg, "Hole mobility in organic single crystals measured by a "flip-crystal" field-effect technique," *Journal Of Applied Physics*, vol. 96, no. 4, pp. 2080–2086, Aug. 2004.

[104] V. Podzorov, E. Menard, A. Borissov, V. Kiryukhin, J. A. Rogers, and M. E. Gershenson, "Intrinsic charge transport on the surface of organic semiconductors," *Physical Review Letters*, vol. 93, no. 8, p. 086602, Aug. 2004.

[105] L. B. Roberson, J. Kowalik, L. M. Tolbert, C. Kloc, R. Zeis, X. L. Chi, R. Fleming, and C. Wilkins, "Pentacene disproportionation during sublimation for field-effect transistors," *Journal Of The American Chemical Society*, vol. 127, no. 9, pp. 3069–3075, Mar. 2005.

[106] C. Reese, W. J. Chung, M. M. Ling, M. Roberts, and Z. N. Bao, "High-performance microscale single-crystal transistors by lithography on an elastomer dielectric," *Applied Physics Letters*, vol. 89, no. 20, p. 202108, Nov. 2006.

[107] O. D. Jurchescu, M. Popinciuc, B. J. van Wees, and T. T. M. Palstra, "Interface-controlled, high-mobility organic transistors," *Advanced Materials*, vol. 19, no. 5, pp. 688–+, Mar. 2007.

[108] G. Horowitz, X. Z. Peng, D. Fichou, and F. Garnier, "Role of the semiconductor insulator interface in the characteristics of pi-conjugated-oligomer-based thin-film transistors," *Synthetic Metals*, vol. 51, no. 1-3, pp. 419–424, Sep. 1992.

[109] C. D. Dimitrakopoulos, A. R. Brown, and A. Pomp, "Molecular beam deposited thin films of pentacene for organic field effect transistor applications," *Journal Of Applied Physics*, vol. 80, no. 4, pp. 2501–2508, Aug. 1996.

[110] H. Klauk, M. Halik, U. Zschieschang, G. Schmid, W. Radlik, and W. Weber, "High-mobility polymer gate dielectric pentacene thin film transistors," *Journal Of Applied Physics*, vol. 92, no. 9, pp. 5259–5263, Nov. 2002.

[111] S. Lee, B. Koo, J. Shin, E. Lee, H. Park, and H. Kim, "Effects of hydroxyl groups in polymeric dielectrics on organic transistor performance," *Applied Physics Letters*, vol. 88, no. 16, p. 162109, Apr. 2006.

[112] A. Assadi, C. Svensson, M. Willander, and O. Inganas, "Field-effect mobility of poly(3-hexylthiophene)," *Applied Physics Letters*, vol. 53, no. 3, pp. 195–197, Jul. 1988.

[113] C. Clarisse, M. T. Riou, M. Gauneau, and M. Lecontellec, "Field-effect transistor with diphthalocyanine thin-film," *Electronics Letters*, vol. 24, no. 11, pp. 674–675, May 1988.

[114] G. Horowitz, D. Fichou, X. Z. Peng, Z. G. Xu, and F. Garnier, "A field-effect transistor based on conjugated alpha-sexithienyl," *Solid State Communications*, vol. 72, no. 4, pp. 381–384, Oct. 1989.

[115] H. Fuchigami, A. Tsumura, and H. Koezuka, "Polythienylenevinylene thin-film-transistor with high carrier mobility," *Applied Physics Letters*, vol. 63, no. 10, pp. 1372–1374, Sep. 1993.

[116] F. Garnier, R. Hajlaoui, A. Yassar, and P. Srivastava, "All-polymer field-effect transistor realized by printing techniques," *Science*, vol. 265, no. 5179, pp. 1684–1686, Sep. 1994.

[117] Z. Bao, A. J. Lovinger, and A. Dodabalapur, "Organic field-effect transistors with high mobility based on copper phthalocyanine," *Applied Physics Letters*, vol. 69, no. 20, pp. 3066–3068, Nov. 1996.

[118] Z. Bao, A. Dodabalapur, and A. J. Lovinger, "Soluble and processable regioregular poly(3-hexylthiophene) for thin film field-effect transistor applications with high mobility," *Applied Physics Letters*, vol. 69, no. 26, pp. 4108–4110, Dec. 1996.

[119] C. D. Dimitrakopoulos, B. K. Furman, T. Graham, S. Hegde, and S. Purushothaman, "Field-effect transistors comprising molecular beam deposited alpha,omega-di-hexyl-hexathienylene and polymeric insulator," *Synthetic Metals*, vol. 92, no. 1, pp. 47–52, Jan. 1998.

[120] H. Sirringhaus, N. Tessler, and R. H. Friend, "Integrated optoelectronic devices based on conjugated polymers," *Science*, vol. 280, no. 5370, pp. 1741–1744, Jun. 1998.

[121] H. E. Katz, A. J. Lovinger, and J. G. Laquindanum, "alpha,omega-dihexylquaterthiophene: A second thin film single-crystal organic semiconductor," *Chemistry Of Materials*, vol. 10, no. 2, pp. 457–+, Feb. 1998.

[122] M. E. Hajlaoui, F. Garnier, L. Hassine, F. Kouki, and H. Bouchriha, "Growth conditions effects on morphology and transport properties of an oligothiophene semiconductor," *Synthetic Metals*, vol. 129, no. 3, pp. 215–220, Aug. 2002.

[123] G. M. Wang, J. Swensen, D. Moses, and A. J. Heeger, "Increased mobility from regioregular poly(3-hexylthiophene) field-effect transistors," *Journal Of Applied Physics*, vol. 93, no. 10, pp. 6137–6141, May 2003.

[124] T. Okuda, S. Shintoh, and N. Terada, "Copper-phthalocyanine field-effect transistor with a low driving voltage," *Journal Of Applied Physics*, vol. 96, no. 6, pp. 3586–3588, Sep. 2004.

[125] I. Mcculloch, M. Heeney, C. Bailey, K. Genevicius, I. Macdonald, M. Shkunov, D. Sparrowe, S. Tierney, R. Wagner, W. M. Zhang, M. L. Chabinyc, R. J. Kline, M. D. Mcgehee, and M. F. Toney, "Liquid-crystalline semiconducting polymers with high charge-carrier mobility," *Nature Materials*, vol. 5, no. 4, pp. 328–333, Apr. 2006.

[126] B. H. Hamadani, D. J. Gundlach, I. McCulloch, and M. Heeney, "Undoped polythiophene field-effect transistors with mobility of 1 cm(2) v-1 s(-1)," *Applied Physics Letters*, vol. 91, no. 24, p. 243512, Dec. 2007.

Acknowledgements

While this thesis bears only one name, many people have contributed to its success. To all of them I would like to express my thanks and gratitude.

First of all I would like to thank Professor Dr. Heinz von Seggern for giving me the opportunity to conduct my doctorate work at his institute. His expertise, openness to new ideas and readiness to discuss as well as his continuous encouragement for improvement of the obtained results have made this thesis possible.

I express my special thanks to Dr. Christian Melzer for the supervision of this thesis. Christian's tireless support for this work, the multitude of critical and constructive discussions, as well as his outstanding and clear view of scientific problems have significantly contributed to the success of this thesis.

Explicit thanks is extended to Professor Dr. Roland Schmechel for supervising the beginning of my Ph.D. work. Roland's willingness to lend an ear during the course of my thesis and his ability to explain complicated topics in simple terms have helped to overcome difficult times.

I especially thank Dr. Thomas Mayer and Eric Mankel for all of their help with the PES experiments, a multitude of fruitful discussions and patient answers to plenty of "short" questions.

I want to highlight the outstanding efforts of "my" students Martin Schidleja, Andrea Gassmann and Tobias Lange, who I worked with during the course of this thesis. I wish to thank all my colleagues at the e-mat for fantastic times. Special thanks go to Marcus Ahles for the excellent introduction to the lab and O-CMOS technology, Oliver Ottinger for endless talks over beers, during bike rides and runs, as well as Christopher Siol for crazy ideas. Further, I want to remember Frederik Neumann, whose life was taken in a tragic accident during my time at the e-mat. I also want to express my gratitude to Gerlinde Dietrich, the retired e-mat secretary, for shedding light into the depths of bureaucracy.

I acknowledge the DFG for its financial support in the framework of the "OFET Schwerpunktprogramm".

Special thanks and gratitude go to Christine Drepper for her support, understanding and patience during hard times, especially during the final phase of my dissertation.

Last but not least I want to thank my parents for their support, their belief in me, their patience and much much more, for which I cannot find the proper words.

Appendix A
List of publications

In conjunction with this work, the following articles were published:

1. N. Benson; "Electronic states at the dielectric / semiconductor interface in organic field effect transistors" in: C. Wöll (ed.); Physical and Chemical Aspects of Organic Electronics; WILEY-VCH 2009.

2. N. Benson, A. Gassmann, E. Mankel, T. Mayer, C. Melzer, R. Schmechel and H. v. Seggern; "The role of Ca traces in the passivation of SiO_2 dielectrics for electron transport in pentacene OFETs"; J. Appl. Phys. 104, 054505 (2008).

3. N. Benson, C. Melzer, R. Schmechel and H. von Seggern; "Electronic states at the dielectric / semiconductor interface in organic field effect transistors"; PSS (a), 205, 3, 475(2008), DOI 10.1002.

4. N. Benson, M. Schidleja, C. Siol, C. Melzer and H. von Seggern; "Dielectric interface modification by UV irradiation: A novel method to control OFET charge carrier transport properties"; Proc. SPIE 6658, 66580W (2007), 10.117/12.733646.

5. N. Benson, M. Schidleja, C. Melzer, R. Schmechel and H. von Seggern; "Complementary organic field effect transistors by ultraviolet dielectric interface modification"; Appl. Phys. Lett 89, 182105 (2006).

6. N. Benson, M. Ahles, M. Schidleja, A. Gassmann, E. Mankel, T. Mayer, C. Melzer, R. Schmechel and H. von Seggern; "Organic CMOS-Technology by Interface Treatment"; Proc. SPIE 6336, 63360S (2006), DOI 10.1117/12.680049.

Appendix B

Additional PES spectra

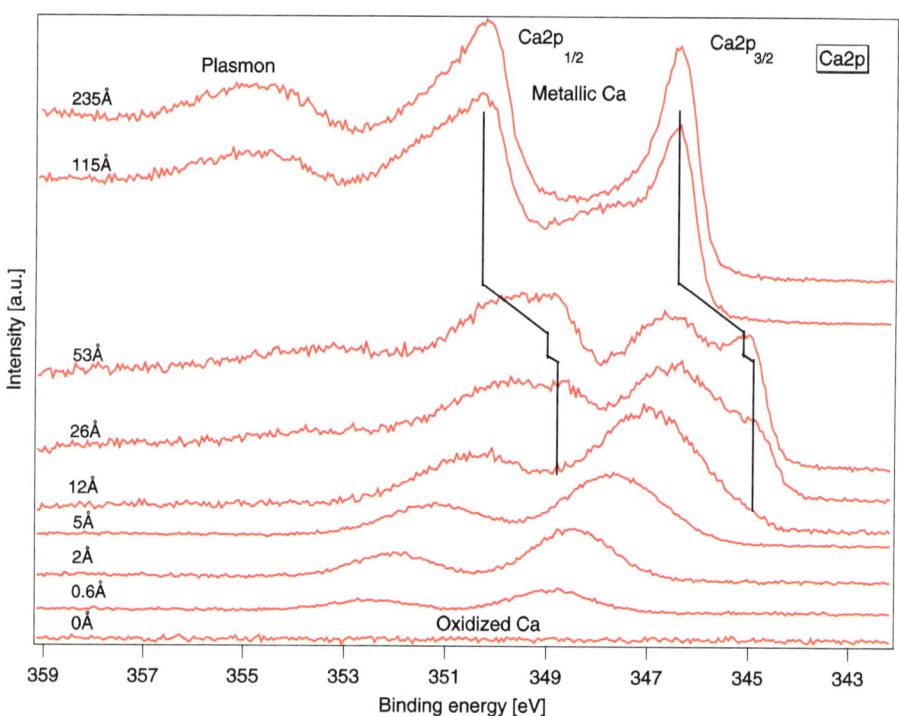

Figure B.1: Ca2p emission spectra on SiO_2 for ascending Ca layer thickness. The Ca layer thickness is enhanced in between the respective measurements without breaking the vacuum.

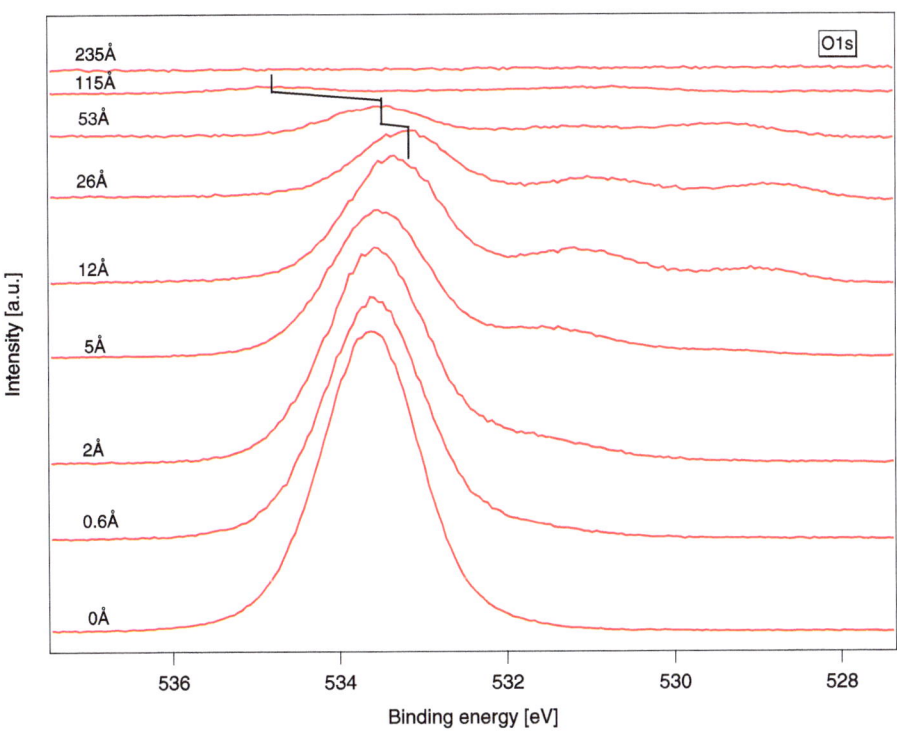

Figure B.2: O1s emission spectra for ascending Ca layer thicknesson SiO$_2$. The Ca layer thickness is enhanced in between the respective measurements without breaking the vacuum.

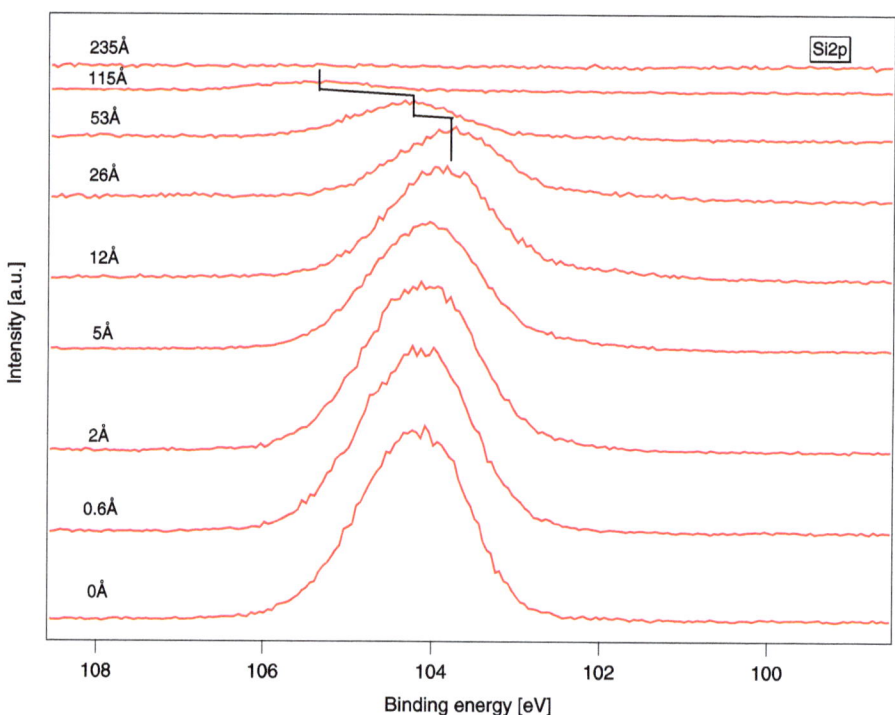

Figure B.3: Si2p emission spectra for ascending Ca layer thickness on SiO$_2$. The Ca layer thickness is enhanced in between the respective measurements without breaking the vacuum.

Appendix C

Development in OFET mobility

Year	Hole mobility $[\frac{cm^2}{Vs}]$	Material Single Crystal	References
2003	0.3	Pentacene	[101]
2003	8	Rubrene	[102]
2004	1.4	Pentacene	[103]
2004	20	Rubrene	[104]
2005	2.2	Pentacene	[105]
2006	19	Rubrene	[106]
2007	40	Pentacene	[107]

Table C.1: Single crystal OFET p-type mobility between 1992-2007 (Figure 1.1).

Year	Pentacene (Thin Film)		References
	Hole mobility $[\frac{cm^2}{Vs}]$	Electron mobility $[\frac{cm^2}{Vs}]$	
1992	0.002	-	[108]
1995	0.0038	-	[109]
1997	1.5	-	[13]
2002	3	-	[110]
2003	-	10^{-6}	[55]
2004	-	0.19	[28]
2006	5.5	-	[111]
2007	-	0.24	[84]

Table C.2: Pentacene OFET mobility between 1992-2007 (Figure 4.1).

Year	Hole mobility $[\frac{cm^2}{Vs}]$	Material Thin Film	References
1984	$1.5*10^{-5}$	Merocyanine	[9]
1986	10^{-5}	Polythiophene	[10]
1988	0.001	Phthalocyanine	[112]
1988	10^{-4}	Poly(3-hexylthiophene)	[113]
1989	10^{-3}	α − sexithiophene	[114]
1992	0.002	Pentacene	[108]
1992	0.027	Oligothiophene	[108]
1993	0.05	$\alpha - \omega -$ dihexyl − sexithiophene	[115]
1994	0.06	Oligothiophene	[116]
1995	0.038	Pentacene	[109]
1996	0.02	Copper phthalocyanine	[117]
1996	0.045	Poly(3-hexylthiophene)	[118]
1997	1.5	Pentacene	[13]
1997	0.13	$\alpha - \omega -$ dihexyl − sexithiophene	[119]
1998	0.1	Poly(3-hexylthiophene)	[120]
1998	0.23	$\alpha - \omega -$ dihexyl − sexithiophene	[121]
2002	3	Pentacene	[110]
2002	0.33	8T-Oligothiophene	[122]
2003	0.2	Poly(3-hexylthiophene)	[123]
2004	0.015	Copper phthalocyanine	[124]
2006	5.5	Pentacene	[111]
2006	0.6	Poly(thieno[3,2-b]thiophene)	[125]
2007	1	Poly(2,5-bis(3-tetradecylthiophen-2yl) thieno[3,2-b]thiophene)	[126]

Table C.3: Development in OFET p-type mobility between 1984 and 2007. For the mobility development of figure 1.1, the various p-type materials are grouped together in families of similar molecular core parts.

Appendix D

Process Parameters

D.1 Thin film deposition from solution

Process parameters

Polymer	PMMA	PC	P4VP	PS	PI
Pre process step	Cleaning with ethanol[1]				
Filter [μm]	0,2		0,02		0,2
Ramp 1	1 s				1 s
Rpm 1	2000				3000
Time 1	1 s				1 s
Ramp 2	1 s				1 s
Rpm 2	3000				4500
Time 2	30 s				60 s
Ramp 3	1 s				1 s
Rpm 3	3000				4500
Time 3	1 s				1 s
Ramp 4	3 s				3 s
Temper step	60 °C for 10 min			-	100 °C for 10 min, 180 °C for 1 h [2]
Layer thickness [nm]	119	132	212	143	169

Table D.1: Process parameters for the polymer deposition in inert N_2. The exact dilutions of the respective polymers are listed in table D.2

Comments:

- [1] The pre process cleaning step is conducted by depositing ethanol on to the substrate. The substrate is dried by using the respective spin coating process.

- [2] For this temper step the substrate is covered with a petri dish

Abbreviations and dilutions

Polymer	Abbreviation	Dilution
Polymethylmethacrylat	PMMA	2% wt in Tetrahydrofuran
Polycarbonat	PC	2% wt in Tetrahydrofuran
Polystyrene	PS	1.7% wt in Toluol
Polyimid	PI	JSR-AL-1054
Poly(4-vinylphenol)	P4VP	2% wt in Tetrahydrofuran

Table D.2: Abbreviations and dilutions of the utilized polymers

D.2 Thin film deposition by PVD

Material	Pentacen	Calcium	Calcium	Gold
Process chamber	OK	MK	OK	MK
Sample position	2	1	2	1
Crucible position	Back right	Front right	Front right	Front left
Process number	11	4	12	1
Transformer	1	5	2	5
Transformer power [%]	19,5	3,5	72-86	8
Sensor	1	3	2	3
Tooling	178,5	59	50	30.5
Rate [Å/s]	2	2	2	2
Sample thickness [nm]	50	100	100	100

Table D.3: Process parameters for evaporated thin films

VDM Verlagsservicegesellschaft mbH

Die VDM Verlagsservicegesellschaft sucht für wissenschaftliche Verlage abgeschlossene und herausragende

Dissertationen, Habilitationen, Diplomarbeiten, Master Theses, Magisterarbeiten usw.

für die kostenlose Publikation als Fachbuch.

Sie verfügen über eine Arbeit, die hohen inhaltlichen und formalen Ansprüchen genügt, und haben Interesse an einer honorarvergüteten Publikation?

Dann senden Sie bitte erste Informationen über sich und Ihre Arbeit per Email an *info@vdm-vsg.de*.

Sie erhalten kurzfristig unser Feedback!

VDM Verlagsservicegesellschaft mbH
Dudweiler Landstr. 99
D - 66123 Saarbrücken

Telefon +49 681 3720 174
Fax +49 681 3720 1749

www.vdm-vsg.de

Die VDM Verlagsservicegesellschaft mbH vertritt

Printed by Books on Demand GmbH, Norderstedt / Germany